The Great Adaptation

The Great Adaptation

Climate, Capitalism and Catastrophe

Romain Felli

Translated by David Broder

VERSO

London • New York

This translation was partially funded by the Faculty of Social and Political Sciences, University of Lausanne (UNIL)

First published in English by Verso 2021
Originally published in French as *La Grande Adaptation. Climat, capitalisme et catastrophe,* Editions du Seuil 2016
Translation © David Broder 2021

1 3 5 7 9 10 8 6 4 2

Verso
UK: 6 Meard Street, London W1F 0EG
US: 20 Jay Street, Suite 1010, Brooklyn, NY 11201
versobooks.com

Verso is the imprint of New Left Books

ISBN-13: 978-1-78873-414-1
ISBN-13: 978-1-78873-417-2 (UK EBK)
ISBN-13: 978-1-78873-416-5 (US EBK)

British Library Cataloguing in Publication Data
A catalogue record for this book is available from the British Library

Library of Congress Cataloging-in-Publication Data
Library of Congress Control Number: 2020948739

Typeset in Minion Pro by Hewer Text UK Ltd, Edinburgh
Printed and bound by CPI Group (UK) Ltd, Croydon CR0 4YY

Contents

Introduction

The Climate: A Problem of Capital Importance

Before he was British ambassador to the United Nations and chef de cabinet to the president of the European Commission, in 1975 Crispin Tickell spent a year's sabbatical at Harvard University. He devoted his time there to studying a serious political problem: climate change. Sir Crispin emphasised the many uncertainties which beset this question – but was nonetheless convinced that human beings had some share of responsibility for the changes now underway. Was this, perhaps, because of the multinationals in the oil industry, or the long history of European colonialism? Not according to the man who would also become an aide to Margaret Thatcher. Rather, the risk lay in the 'increasing temptation for governments', under pressure from the exorbitant demands of the 'hungry, the deprived, or the unemployed, to take actions which not only beggar thy neighbours, but also damage or alter the delicate mechanisms of the atmosphere'.[1]

Two centuries earlier, at the height of the Industrial Revolution, Reverend Thomas Malthus had blamed destitution and famine on the poor (considered too many, given the limits of nature) and on aid to the poor (which was similarly deemed too generous). Sir Crispin continued this neo-Malthusian line of argument. Too destitute, the poor destroyed the environment by over-exploiting natural resources. Too well-organised, they obtained an improper redistribution of wealth from the welfare state. Private property was in danger – and

---

1   Crispin Tickell, *Climatic Change and World Affairs*, Oxford: Pergamon Press, 1977, p. 48.

the state was becoming ungovernable. Or so claimed the conservative thinkers of the time.[2] At Nature's mighty feast, Malthus wrote, there are no vacant covers.

Human overpopulation thus provoked climatic imbalances. But, worse, according to Sir Crispin, it prevented our species from *adapting* to these changes. When the Earth had been less populated, the consequences of climatic variations were easily resolved: migration was a sufficient response. But the demographic explosion – what biologist Paul Ehrlich called the 'population bomb'[3] – rendered this solution impracticable: for whoever did migrate would find others in their way. Overpopulation destroyed the flexibility necessary for adaptation.[4] The result? Conflicts between rich and poor, between North and South, and between present and future.

Happily, in 1982, the famous US environmentalist couple Amory and Hunter Lovins came along to reassure us: climate change ain't gonna happen! For them, it was pointless to worry about the concentration of carbon dioxide in the atmosphere, for the market was going to solve the problem anyway. As they put it, 'Economic rationality . . . can be our strongest *ally* in curbing the worst assaults on the climate.'[5] After all, the Lovinses explained, in a market economy, it was impossible that people would go on using fossil fuels that harmed the climate (petrol, coal, gas, etc.): for this would be too expensive.

Fast forward to today, and oil prices are at rock-bottom and global warming is sky-high . . . And this despite three decades of constant extension of the 'free market', as per the neoliberal utopia. No problem! If the market has failed to reduce greenhouse gas emissions, at least it can help us manage the climate disasters – or so claims

---

2 Michel Crozier, Samuel Huntington and Joji Watanuki, *The Crisis of Democracy: Report on the Governability of Democracies to the Trilateral Commission*, New York: New York University Press, 1975. Tickell spent his year at Harvard in Huntington's department, and it was Huntington who edited his book.

3 Paul R. Ehrlich, *The Population Bomb*, New York: Ballantine Books, 1968.

4 Tickell, *Climatic Change*, p. 38.

5 Amory B. Lovins and L. Hunter Lovins, 'Energy, Economics and Climate – an Editorial', *Climatic Change*, vol. 4, 1982, p. 220; author's emphasis.

a certain Hunter Lovins. The 'emerging adaptation market', she notes, is rich in opportunities for products like flood risk prevention measures, the preparation of aid shelters and emergency communications services, and prevention and protection against fires. The climate crisis allows for the extension of the market's grip: for 'wherever there is a climate problem there are climate capitalists looking for solutions.' In short, it's 'good business'.[6]

This book tells the story of how the idea of adapting to climate change has been mobilised – and deployed – since the 1970s for the purpose of allowing the market to reach into all domains of life. The introduction that follows, which probably has a sharper edge than the book as a whole, explains the theory which is the bedrock of my analysis.

THE ANTHROPOCENE'S LACK OF CLASS

Thanks to the combined work of the social sciences, natural sciences, environmental historians – and activists, especially in the Global South – we know that the causes and consequences of global warming are not shared out equally.[7] This inequality is, moreover, at the basis of international climate law, with the principle of 'common but differentiated responsibilities'. Seeking to explain the origins of the problem, some point to the role of rich countries, industry, the oil sector, the consumerist way of life, or even Western civilisation itself. Others foreground its consequences for the poorest, the Global South, poor farmers, or ecosystems. More generally, global warming is integral to the grand narrative of the Anthropocene, an analysis holding that with the Industrial Revolution, human society became a geological force properly speaking, on account of its activities as well as its growing population. Fundamentally altering the functioning of the great systems

6   Hunter Lovins and Boyd Cohen, *Climate Capitalism: Capitalism in the Age of Climate Change*, New York: Hill and Wang, 2011, quotes from pp. 250, 265 and 269.

7   J. Timmons Roberts and Bradley Parks, *A Climate of Injustice: Global Inequality, North–South Politics, and Climate Policy*, Cambridge, MA: MIT Press, 2006.

of nature at a now planetary scale, human society is, in this view, provoking extremely dramatic consequences.[8]

But, in blaming humanity as a whole for the transformations taking place in a nature portrayed as standing outside of society, the concept of the Anthropocene risks obscuring things more than it clarifies them. The major transformations in the fabric of life – a fabric which inextricably binds relations, processes and materials, which are simultaneously biophysical, organic, material and cultural – are very real.[9] Unleashed already in the sixteenth century, these transformations preceded and informed the Industrial Revolution. Far from being the result of humanity's predatory action against nature, they constitute a particular way of organising the economy, which is also a particular way of organising nature. Indeed, the 'nature' thus produced (not *ex nihilo*, but through the working and remodelling of pre-existing elements) is itself historical, just as history is natural. But to speak of a 'production' of nature does not imply that the nature thus produced is mastered or controlled by humanity or subjected to it – quite the contrary!

This *historical* nature is the crystallisation of the fabric of contemporary life. It is the result of a long and global history; it consists of

---

8   For a – critical – contextualisation of this idea and a historical synthesis of the breadth of transformations, see Christophe Bonneuil and Jean-Baptiste Fressoz, *The Shock of the Anthropocene: The Earth, History and Us*, London: Verso, 2016.

9   Here, I am (all too rapidly) summarising a way of conceptualising this history which takes seriously the contributions of environmental history, political economy and the natural sciences. This approach tries to develop a language which allows us to go beyond the pitfalls of the society/nature dualism that deeply imbues contemporary thought and especially ecologist thought – though it does not completely succeed in this regard. Jason Moore's very ambitious project certainly represents one of the most important advances in this direction: see Jason W. Moore, *Capitalism in the Web of Life: Ecology and the Accumulation of Capital*, London and New York: Verso, 2015. The concept of 'fabric of life' I propose here is an attempt to translate, in conjunction, the ideas of web of life and of *oikeios* advanced by Moore. Moore builds his analysis on authors important to my line of argument like David Harvey, *The Limits to Capital*, London and New York: Verso, 2018, and *Justice, Nature, and the Geography of Difference*, Oxford: Blackwell, 1996, and Neil Smith, *Uneven Development: Nature, Capital, and the Production of Space*, 3rd edn, Athens: University of Georgia Press, 2008 [1984]. This latter also established the concept of the 'production of nature', which he elaborated basing himself on Henri Lefebvre, *The Production of Space*, Oxford: Blackwell, 2009 [1974].

power relations, the displacement of matter, energy and biological species, of social relations, of slavery, of exploitation, of cultures, of economic accumulation, etc. And all this did not take place completely by chance. If, for example, the geography of oil extraction does not correspond to the geography of its use, the reasons for this are far from exclusively geological.[10]

The replacement of renewable energies (wind, water . . .) by coal at the beginning of the industrial revolution did not owe to some simple technological advantage that coal had over renewables. Rather, it was used the better to subject the uprooted peasants to the industrial machinery which was producing for the market.[11] And, if the contemporary urban ecology, especially in North America, is dominated by the individual car – and all that goes with it, from pollution to the squandering of resources and the boundless sprawl of the suburbs – this is because oil companies, auto manufacturers, credit institutions and the US state have produced the infrastructure necessary for the development of this particular 'way of life'. It is not as if it was coded into Americans' DNA.[12] Particular social relations (class relationships) organise the transformations in the fabric of life, just as this fabric – as far as humans are concerned, anyhow – exists by way of the relations which it itself makes possible.

The study of the longer history of the relations that make up the fabric of life, and have done so since the sixteenth century, must pass by way of the history of capitalism. 'Capitalism is not an economic

10   Claude Raffestin, *Pour une géographie du pouvoir*, Paris: Litec, 1980.

11   Andreas Malm, 'The Origins of Fossil Capital: From Water to Steam in the British Cotton Industry', *Historical Materialism*, vol. 21, no. 1, 2013, pp. 15–68. The deep-rooted relationship between the control/exploitation of workers and the twentieth-century development of fossil fuels, especially petrol, has been explored by Timothy Mitchell in his *Carbon Democracy: Political Power in the Age of Oil*, London and New York: Verso, 2011.

12   Matthew T. Huber, *Lifeblood: Oil, Freedom, and the Forces of Capital*, Minneapolis: University of Minnesota Press, 2013. Analyses by André Gorz and Ivan Illich on the social ideology of the car already pointed in this direction. In a quite different style – and from a reactionary point of view – I recommend the chapter entitled 'La bagnole et l'arbre' by Louis Chevallier in *L'Assassinat de Paris*, Paris: Éditions Ivrea, 1997 [1977].

system; it is not a social system; it is *a way of organizing nature*.[13] We do not live in the Anthropocene, but in the Capitalocene.[14]

THE CLIMATIC THREAT

Whoever wants to talk about global warming cannot, then, just stick to CO_2 emissions, the role of the oil industry or the place of the car in our societies (crucial though these elements are). In embracing the illusion that the sole variable 'CO_2' is the only source of the problem, we act as if we could control and limit the climate problem – or even make it disappear.[15] But more than greenhouse gas emissions, what is at stake in the climate question is the particular way in which nature is organised.

Despite their apparent differences, one thing Tickell's neo-Malthusianism and Lovins's neoliberalism have in common is that they speak not just about nature, the environment and the climate, but also about social fears, market mechanisms and ways of producing the environment. They tell us that the climate question – the problem of climate change owing to human activity, in particular linked to greenhouse gas emissions – is a political question. In painting the hungry, the deprived, or the unemployed as the problem, on account of their excessive numbers, or capitalist entrepreneurs as the solution, they propose a class analysis and class solutions to global warming.

As we shall see in chapter 2, research on climate change since the 1970s has consistently been imbricated with the development of political projects to respond to it. Some elites, especially in the United States – at the point where the atmospheric and agronomic sciences

---

13   Moore, *Capitalism in the Web of Life*, p. 2, his emphasis.

14   This concept is proposed by Moore. See also Christophe Bonneuil and Jean-Baptiste Fressoz, *The Shock of the Anthropocene*, and Andreas Malm and Alf Hornborg, 'The Geology of Mankind? A Critique of the Anthropocene Narrative', *Anthropocene Review*, vol. 1, no. 1, 2014, pp. 62–9.

15   Erik Swyngedouw, 'Apocalypse Forever? Post-political Populism and the Spectre of Climate Change', *Theory, Culture & Society*, vol. 27, nos. 2–3, 2010, pp. 21–232.

cross paths with senior administration, big private foundations (Ford, Rockefeller, etc.) and businesses – take the different problems related to humanity's involuntary alteration of the climate as existential problems for capitalism's very survival.

And these are, indeed, problems, at two levels. Firstly, any drastic reduction in greenhouse gas emissions would mean changing the economic organisation of capitalism based on fossil fuels. It would imply attacking the very bedrock of the profits of the most powerful US firms, those in the energy and auto sectors. That is the main reason why these companies, backed by many states, are doing as much as they can to put the brakes on the enactment of a climate policy that would reduce greenhouse gas emissions.[16] That is also why the effective reduction of these greenhouse gas emissions seems impossible.

But climate change threatens capitalism at a second level, namely, in terms of a crisis of social reproduction. Hurting agricultural productivity (through more frequent droughts), exposing habitations to rising water levels and destroying public infrastructure through the proliferation of extreme climate events, global warming pushes up the cost of social reproduction (food, housing, energy, etc.). And this cost will potentially have further repercussions by raising the cost of the workforce for capital – whether directly, in terms of increased wages, or indirectly, through the taxes and charges necessary to provide public infrastructure.[17] This would mean the end of the 'low-cost nature' which has characterised capitalist accumulation since the sixteenth century . . . We should thus return to the usual formulation of ecological crisis. As Moore writes, 'The crises of capitalism-in-nature are

16   Andriana Vlachou, 'Capitalism and Ecological Sustainability: The Shaping of Environmental Policies', *Review of International Political Economy*, vol. 11, no. 5, 2004, pp. 926–52; Peter Newell and Matthew Paterson, *Climate Capitalism: Global Warming and the Transformation of the Global Economy*, Cambridge, Cambridge University Press, 2010; Romain Felli, 'Environment, Not Planning: The Neoliberal Depoliticisation of Environmental Policy by Means of Emissions Trading', *Environmental Politics*, vol. 24, no. 5, pp. 641–60.

17   James O'Connor, *Natural Causes*, New York: Guilford Press, 1998. For a preliminary application of this reading to the climate, see Romain Felli, 'On Climate Rent', *Historical Materialism*, vol. 22, nos. 3–4, 2014, pp. 251–80.

crises of what nature *does for* capitalism, rather more than what capitalism *does to* nature'![18]

The climate crisis is a crisis in the capitalist way of organising nature. The young law professor Richard Falk posed the key question in the clearest of terms when he asked, already back in 1971, 'Can the capitalist system and the market economy be adapted to the needs of the ecological age?'[19]

One response to this question – perhaps the most important question – has consisted of denying the existence of global warming, minimising its importance or sowing doubt over the results of scientific inquiry.[20] But as useful as this tactic may be for some companies as a means of buying time, it quite obviously does not resolve the fundamental problems of social reproduction.

Since the 1970s, one section of elites, who could be termed 'enlightened reactionaries',[21] have sought to formulate a more serious set of responses. These responses take the reality of the ecological and climatic crisis on board, but they are simultaneously meant to safeguard the pursuit of economic accumulation in the long term. These elites are perfectly conscious that economic accumulation poses a problem, both as a drain on natural resources and as a factor in increased pollution. Their solutions propose (in words at least – the practice is quite a different story!) to reduce the environmental 'impact' of production. This means minimising greenhouse gas emissions, a relative reduction in

---

18 Moore, *Capitalism in the Web of Life*, p. 27; his emphasis.

19 Richard A. Falk, *This Endangered Planet: Prospects and Proposals for Human Survival*, New York: Random House, 1971, p. 360.

20 Naomi Oreskes and Erik M. Conway, *Merchants of Doubt: How a Handful of Scientists Obscured the Truth on Issues from Tobacco Smoke to Global Warming*, London: Bloomsbury, 2012.

21 Karl Polanyi used the term 'enlightened reactionaries' to refer to that fraction of the nineteenth-century British aristocracy which backed the limiting of the working day – less for humanitarian reasons than to defend its own interests. Karl Polanyi, *The Great Transformation*, Boston: Beacon, 2014 [1944].

the quantity of energy and materials used per unit of output, etc. They also rely on the idea of finding narrowly technological remedies to the problem (such as geo-engineering).

The promotion of certain economic sectors like renewable energies on the pretext that they are better for the climate has driven the formation of a class fraction of 'green' capitalists. In a similar vein, it is argued that companies reducing their environmental impact (through eco-efficiency, energy-saving, etc.) would make savings in terms of production costs. And, given that regulations are now inevitable, one additional response has been to try to water down these regulations, or make sure that they operate according to a market logic (through a market in pollution permits, or ecotaxes, or voluntary standards . . .) in order to minimise their impact on how companies operate.[22]

Thus, a worldview that saw economic and demographic growth as the source of environmental problems gave way, first, to a theory which asserted the compatibility between the environment and economic growth (sustainable development) and then, soon after that, to a political strategy that posed the expansion of the market as the solution to environmental problems. With this move came the advent of a liberal environmentalism.[23]

---

22   These last two responses form the basis of strategies for the so-called 'ecological modernisation' of capital. There is a very abundant literature on this score: I especially recommend Maarten A. Hajer, *The Politics of Environmental Discourse: Ecological Modernization and the Policy Process*, Oxford: Oxford University Press, 1995; and, on the climate, Newell and Paterson, *Climate Capitalism*; Daniel Tanuro, *Green Capitalism: Why it Can't Work*, Halifax: Fernwood, 2014; Larry Lohmann, *Carbon Trading: A Critical Conversation on Climate Change, Privatisation and Power*, Uppsala: Dag Hammarskjöld Foundation, 2006.

23   Steven F. Bernstein, *The Compromise of Liberal Environmentalism*, New York: Columbia University Press, 2001. I provide a summary of these theses and apply them to the emergence of the problematic of adaptation in Romain Felli, 'Adaptation et résilience: critique de la nouvelle éthique de la politique environnementale internationale', *Éthique publique*, vol. 16, no. 1, 2014, pp. 101–20. See also Dominique Pestre, 'Néolibéralisme et gouvernement', in Dominique Pestre (ed.), *Le Gouvernement des technosciences: Gouverner le progrès et ses dégâts depuis 1945*, Paris: La Découverte, 2014, pp. 261–84.

THE GREAT ADAPTATION

Less visible – but, I would argue, just as important – is a parallel strategy, elaborated since the 1970s, to respond to the crises in the capitalist production of nature. This strategy has sought to remodel socio-ecological relations – the fabric of life – in order to adapt them to the multiple crises in this production. This allows the debate on the climate crisis to be shifted away from the question of reducing greenhouse gases and toward the transformation of socio-ecological systems, in order to *adapt* them to a changing climate. This means a quest not to avoid the change, but rather to minimise its consequences. It can even mean embracing climate change in order to draw profits from it. As we shall see below, the solution that has been employed in order to make societies more flexible, more reactive, and more adaptable to climate change is the market. The 'great adaptation' answers the climate crisis not by reining in the market, but by expanding it.

This name makes reference to *The Great Transformation*, the 1944 work by the anthropologist, economist and historian Karl Polanyi.[24] Writing in the apocalyptic times that were the Second World War, the socialist Polanyi cast the development of that liberal utopia, the free market, as the cause of the catastrophe of his day. Rediscovering Marx's analysis of 'primitive accumulation' – albeit not always crediting him for it – Polanyi analysed the conditions in which the bourgeois class had attempted to create the liberal utopia of a market society between the seventeenth and nineteenth centuries. In place of the integration of economic, social and cultural activities – bound through ties of dependence, reciprocity and attachment to the land – the liberals had tried to separate the economy and society, and partly succeeded in this. But far from freeing up some natural human propensity to truck, barter, and exchange, the great transformation itself had to build an economic rationality. The peasants were dispossessed from their common lands by the

---

24   Polanyi, *The Great Transformation*.

all-too-visible hand of the state. As Polanyi put it, 'laissez-faire was planned'. The revision of the Poor Laws in Britain, which had offered some minimal support to the most destitute through the parishes, helped to create a precarious workforce. Thus made 'free', the dispossessed peasants were forced to sell their capacity to work to the capitalists in order to survive. The land of which they were thus dispossessed began to circulate in the economy as a simple commodity, which assumed an economic price and could be accessed in exchange for payment.

The free, self-regulating market thus required the creation of two 'fictitious' commodities: labour and land. These commodities are fictitious because though they do, indeed, have a price (the wage, or rent, respectively) they cannot be produced in market conditions. The fictitious character of these commodities thus dooms to failure the liberal utopia of the maximal extension of the market . And in the meantime, this liberal utopia also destroys the protections that society had constructed for itself. In transforming nature and human beings into commodities, it destroys the very bases of life. And it prompts 'counter-movements' which seek the protection of society: from trade unions seeking to decommodify labour-power to landowners who demand protectionist tariffs at the borders.

Polanyi's analysis was subsequently borne out by the building of the welfare state after the Second World War. The combined development, in the industrialised countries, of both the power of trade unions and of socialist or social-democratic parties, as well as feminist, environmentalist and anti-racist social movements, helped to rebuild the forms of social protection that had been destroyed by the prewar liberal utopia. The 'Trente Glorieuses' or Golden Age that followed 1945 should not be idealised – the social protections thus constructed have often been exclusive, reserved to men and or to citizens, and based on the perpetuation of a colonial regime and an untrammelled exploitation of natural resources. But these years doubtless constituted a moment of re-regulation of capitalism, restraining capital's freedom of action and socialising a growing part of social reproduction.

These restrictions on investors' freedom were the very thing that the neoliberals went on the assault against from the late 1970s on. Seeking to restore companies' profits and discipline a workforce that had grown too independent, neoliberalism – a capitalist class project – systematically attacked these protections for society.[25] Trade unions, the peoples of the Global South and environmental protection were the three main targets of the neoliberal movement to re-extend the market. We need to take this 'counter-counter-movement' into account if we want to understand the contemporary reorganisations of the capitalist production of nature – as well as its implications for the climate crisis. As Naomi Klein aptly notes, the neoliberal movement arrived at the worst possible moment, from the viewpoint of the climate crisis. At the very moment that we need a strong regulation of the economy, and states which have the capacity to plan, we find that the neoliberals have systematically destroyed these same capacities.[26]

But the neoliberal movement has done a lot more than that. This book argues that environmental questions and the climate question, in particular, are far from merely peripheral objects of the neoliberal project. Rather, they have been important to the redefinition of a capitalist nature. Neoliberalism is a way of organising nature – and the Great Adaptation is part of the political project of those who have sought to orchestrate this reorganisation. Promoting flexibility and adaptability (as we shall see in chapter 2) and establishing market rationality as the adaptive response to the transformations of the climate (chapters 3 and 4), the Great Adaptation seeks to respond to the climate crisis with further steps to integrate populations, ecosystems and institutions into capital accumulation.

---

25   David Harvey, *A Brief History of Neoliberalism*, Oxford: Oxford University Press, 2005; David McNally, *Global Slump: The Economics and Politics of Crisis and Resistance*, Oakland, CA: PM Press, 2010; Leo Panitch and Sam Gindin, *The Making of Global Capitalism: The Political Economy of American Empire*, London and New York: Verso, 2012.

26   Naomi Klein, *This Changes Everything: Capitalism vs. the Climate*, London: Penguin, 2014.

THE ADAPTATIONIST INTERNATIONAL

This book does not revisit the debates on the 'international climate regime' and the problems of adaptation posed therein.[27] It is not that these are questions of merely secondary importance. Truth be told, one could reasonably argue that the current climate regime – since the Copenhagen summit in 2009, in any case – no longer truly seeks to reduce greenhouse gas emissions. The pseudo-success of the 2015 Paris Agreement is emblematic of this impasse. Despite its general- ised acknowledgement of the problem, there was no mention there of binding measures to reduce greenhouse gas emissions. In this sense, we would rightly speak of a 'turn toward adaptation' in international climate policy, even though it is a turn by default: for adapting is the only choice left when you have given up on anything else.

Having abandoned the objective of reducing or controlling global warming, the international climate regime's real focus today is to find some way of compensating for the wrongs done to the Global South by global warming. And the language chosen for this operation is the language of adaptation – for this makes it possible to avoid any reference to rich countries' own responsibility. In helping the countries of the South build up their 'capacities for adaptation', the Northern countries pretend to be offering a humanitarian service, not acknowledging that they them- selves have any debt to repay. But the help provided in the name of adap- tation remains . . . essentially a matter of nice-sounding words. Despite reiterated promises, and in particular the $100 billion a year promised at Copenhagen, the various adaptation funds dedicated to the Global South remain desperately under-resourced, controlled by the donor countries and in competition with other forms of economic 'aid'.[28] Added to that, as

---

27 I have no hesitation in making this choice, since any readers so interested can instead refer to Stefan C. Aykut and Amy Dahan, *Gouverner le climat?: Vingt ans de négociations internationales*, Paris: Presses de Sciences Po, 2015, which covers these questions in some detail.

28 David Ciplet, J. Timmons Roberts and Mizan Khan, 'The Politics of International Climate Adaptation Funding: Justice and Divisions in the Greenhouse', *Global Environmental Politics*, vol. 13, no. 1, 2013, pp. 49–68.

we shall see in chapters 3 and 4, we may well ask if the programmes for adaptation to climate change that have actually been rolled out seek to reduce the vulnerability of the populations affected or, more prosaically, to increase their adaptation to global capitalism . . .

This book analyses the emergence of adaptation as a political problem, and some facets of its use since the 1970s. In so doing, I hope that it can also contribute to the understanding of the geopolitics of global warming,[29] of the history of the science of climate change and of the failures of the 'international climate regime' – that set of more or less formal rules and accords (like the United Nations Framework Convention on Climate Change, the Kyoto Protocol and the Paris Agreement) which seek to govern the global climate.[30] It can also inform the study of national and international policies for adapting to climate change.[31]

UNDER THE SKY, IDEAS

Since the field of climate crisis first arose, scientific and political ideas have always been inextricably linked. It is generally accepted that political proposals to manage the climate crisis should be based on scientific expertise. But we should also turn that proposition around and admit that the scientific ways of seeing, investigating and explaining the problem themselves bear a certain vision of the world.

29 François Gemenne, *Géopolitique du changement climatique*, Paris: Armand Colin, 2009.

30 Aykut and Dahan's *Gouverner le climat?* is the work of reference on this question.

31 I think the best French-language contributions on this subject are those in Valentine Van Gameren, Romain Weikmans and Edwin Zaccai, *L'Adaptation au changement climatique*, Paris: La Decouverte, 'Reperes', 2014. In English, Mark Pelling, *Adaptation to Climate Change: From Resilience to Transformation*, London: Routledge, 2011; and Marcus Taylor, *The Political Ecology of Climate Change Adaptation: Livelihoods, Agrarian Change and the Conflicts of Development*, London: Routledge, 2015. Johann Dupuis (*S'adapter au changement climatique: Analyse critique des nouvelles politiques de gestion de l'environnement. Cas spécifiques de l'agriculture en Inde et du tourisme hivernal en Suisse*, Neuchâtel: Alphil, 2015) shows how the adjective 'adaptive' has justified the rollout – and financing – of infrastructure projects that respond to pre-existing economic or political imperatives as much as they do to climate change.

For instance, the development of our atmospheric understanding relies on complex weather-data collection networks, institutional equipment and scientific models, which are in part a legacy of the questions – and the financing – of the Cold War era.[32] In this regard, the sociologist of sciences Sheila Jasanoff has spoken of a 'co-production of science and social order' – an argument which Clark Miller has explicitly grounded with specific relation to the climate.[33] The history of climatology and the sociology of the climate sciences are rapidly growing fields, and it is not my goal to offer any summary of their results.[34] But they allow us to pay closer attention to the intrinsically political dimension of the questions of adaptation to climate change. Conceptualisations of adaptation to climate change are not walled off in some academic cloister: rather, they constantly irrigate the practice, policies and financing of such projects, and introduce new inequalities and new forms of domination. That is why an examination of how they arise, how they are used, and what their effects are, is so important. This book's aim is to contribute to that investigation.

32   Paul N. Edwards, *A Vast Machine: Computer Models, Climate Data, and the Politics of Global Warming*, Cambridge, MA: MIT Press, 2010.

33   Clark A. Miller, 'Climate Science and the Making of a Global Political Order', in Sheila Jasanoff (ed.), *States of Knowledge: The Coproduction of Science and Social Order*, New York: Routledge, 2004.

34   It is worth consulting Clark A. Miller and Paul N. Edwards (eds), *Changing the Atmosphere: Expert Knowledge and Environmental Governance*, Cambridge, MA: MIT Press, 2001; Spencer R. Weart, *The Discovery of Global Warming*, Cambridge, MA: Harvard University Press, 2008; Joshua P. Howe, *Behind the Curve: Science and the Politics of Global Warming*, Seattle: University of Washington Press, 2014; James Rodger Fleming, *Fixing the Sky: The Checkered History of Weather and Climate Control*, New York: Columbia University Press, 2010; Mike Hulme, *Why We Disagree About Climate Change: Understanding Controversy, Inaction and Opportunity*, Cambridge: Cambridge University Press, 2009.

Chapter 1

Climate Crisis and Capitalism's Survival

At the beginning of the 1970s, the intellectuals, scientists and politicians pondering the future of Western societies – and especially the United States – were obsessed by the question of survival. The crisis of morale provoked by the war in Vietnam as well as the spread of social conflicts raised doubts about the future of the 'free enterprise system'. Their fears over capitalism's survival crossed paths with fears over the survival of the planet itself. Environmentalist sentiments were, at that time, expressed in a catastrophist form which used both a Biblical imagery – metaphors like the Flood, the Apocalypse or Genesis – and a vision of human society reduced to its biological dimension. For the science historian Jacob Hamblin, 'Much of what is today considered pro-environment literature, in the 1960s and 1970s, was in fact human survival literature.'[1]

We would struggle to imagine the level of vulnerability that these intellectuals attributed to capitalism at the turn of the 1960s and 1970s. For them, this mode of production's very survival was at stake, especially in the United States. This context shaped the way in which the environmental crisis was to be handled, indeed at a global scale. A growing part of the population challenged the existing order, organised to win new rights and no longer recognised the state or the economic system as having moral authority. The Civil Rights struggles in the United States, the feminist, workers' and anti-racist

---

1 Jacob Darwin Hamblin, *Arming Mother Nature: The Birth of Catastrophic Environmentalism*, Oxford: Oxford University Press, 2013, p. 165.

movements, and the decolonisation struggles waged by the peoples of the Third World mutually reinforced one another. The USSR could still appear to some as a viable counter-model, especially after the end of the Stalinist period. In 1975, *Time* magazine ran the cover title 'Can Capitalism Survive?'[2] Conservative theorists like Ralf Dahrendorf and Raymond Aron worried over the disenchantment with progress. And Jürgen Habermas spoke of a legitimation crisis for capitalism.[3]

NEO-MALTHUSIANISM AND POLITICAL CONSERVATISM

The first *Earth Day* was organised in the United States on 22 April 1970. It brought apocalyptic worries over the state of the planet and the exhaustion of natural resources together with the critique of consumer society.[4] Above all, it legitimised the idea that environmental problems are, above all, problems of human overpopulation, relative to the resources available and the planet's own limits.

This horror at the overpopulation of the world has been common among many environmentalists, students of ecology and biologists, but also economists, politicians and activists who have discussed the environment since the end of the Second World War.[5] Whatever their background or the particular focus of their interests, the US environmentalists of the 1940s and 1950s, from Fairfield Osborne to William Vogt passing via Aldo Leopold, lamented natural systems' limited 'load capacity'. They worried over the growing pressure to which 'man' was subjecting these systems on account of growing populations and rising consumption.

This Malthusian-environmentalist moment in the years between 1946 and 1975 crossed paths with another moment in the domain of

---

2   Cited in Panitch and Gindin, *The Making of Global Capitalism*.
3   Jürgen Habermas, *Legitimation Crisis*, Cambridge: Polity, 1976.
4   Adam Rome, *The Genius of Earth Day: How a 1970 Teach-In Unexpectedly Made the First Green Generation*, New York: Hill and Wang, 2013.
5   Thomas Robertson, *The Malthusian Moment: Global Population Growth and the Birth of American Environmentalism*, New Brunswick, NJ: Rutgers University Press, 2012; and Björn-Ola Linner, *The Return of Malthus: Environmentalism and Postwar Population-Resource Crises*, Cambridge: White Horse Press, 2003.

economics. This second moment gripped the United States as it sought to manage a world on the road to decolonisation. Peoples' struggles for their national independence transformed the relations between core and periphery. As the independent countries asserted their sovereignty over natural resources, the metropoles' access to these goods – hitherto guaranteed by the colonial system – was no longer certain. The US superpower was, therefore, pressured to secure its long-term control over its supply lines and to promote the global expansion of US firms. It was at the same time confronted with mounting Soviet influence among the peoples seeking to liberate themselves. One response to these new constraints took the form of vast projects that sought to combat hunger and poverty in the so-called 'underdeveloped' world. On 20 January 1949, the fourth point in President Truman's inauguration speech thus affirmed the USA's intent to help poor countries to 'develop'.[6]

Nonetheless, as far as elites in the capitalist world were concerned, there was no question of recognising that poverty might be the result of colonialism and exploitation. As in the times of Reverend Malthus, the biologisation of the social would serve as a substitute for political economy. If the poor were poor, the main reason was that there were too many of them. The poor were producing too many offspring relative to the resources available to them, and thus entrenching themselves in poverty. Overpopulation provided a convenient explanation for an unjust situation. But it also represented a threat, for it was creating masses of poor people, hungry and radicalised, ready to come and break down the doors of the Empire.

Picking up on the neo-Malthusian vulgate, texts produced by the US Army warned against the effects of overpopulation. The report by the RAND Corporation's Theodore von Karman drawn up for the US Air Force at the National Academy of Sciences in 1958 warned against the global rise of 'non-whites'. Von Karman detailed the prospect of

---

6   Isabelle Hillenkamp and Jean-Michel Servet, 'La Lutte contre la pauvreté, un enjeu international', *CERISCOPE Pauvreté*, 2012 [online], ceriscope.sciences-po.fr.

coming famines as well as the vulnerabilities of the US owing to its growing dependence on complex systems.[7]

The grammar of overpopulation continued to give form to naturalists' discourse throughout the 1970s and up till the mid 1980s. Biologists like Paul Ehrlich and Garett Hardin, the climatologist and oceanographer Roger Revelle and the physicist James Lovelock counted among the popularisers of this mode of thinking, notwithstanding their sometimes very different political perspectives. The historian Thomas Robertson defends the 'Malthusian environmentalists' (preoccupied with the pressures that population growth would place on the environment) from the worst accusations of racism, sexism and colonialism levelled against them. According to Robertson, the critique of economic growth had animated these writers just as much as the critique of overpopulation – and it blinded them to the political consequences of their theories.[8]

Until the 1970s, in North America environmental protection was a politically conservative endeavour. The creation of natural parks and the preservation of a supposedly intact nature were central to the construction of the American nation. They made up part of the violent dispossession of the native populations' lands. The landscapes which Europeans saw as 'untamed' or 'virgin' territories (metaphors well expressing a patriarchal viewpoint) were in reality the product of a long-term ecological interaction between the original inhabitants and their environment.[9] Much more than that, the conservation of nature as a resource was an important political project promoted by conservative Republicans from the 1900s (Theodore Roosevelt) through the creation of the Resources for the Future (RFF) think-tank in the 1950s.[10] The best stewardship of natural capital would allow

7   Hamblin, *Arming Mother Nature*, pp. 156–7.

8   Robertson, *Malthusian Moment*, p. xv.

9   William Cronon, *Changes in the Land: Indians, Colonists, and the Ecology of New England*, New York: Hill & Wang, 1983.

10   Yannick Mahrane and Christophe Bonneuil, 'Gouverner la biosphère: De l'environnement de la Guerre froide à l'environnement néolibéral', in Pestre (ed.), *Le Gouvernement des technosciences*.

production to draw on natural resources into the long term. In the 1960s and 1970s the potential exhaustion of natural resources again haunted conservatives, for example, a young Texan businessman from a wealthy East Coast patrician family, George H.W. Bush. Entering politics in 1969 he became the president of Young Republicans for the Environment. He fought for limits on birth rates in order to preserve natural resources: as he saw it, this was 'the most critical problem facing the world in the remainder of this century'.[11]

AGRICULTURAL PRODUCTION AND FAMINE

Neo-Malthusians constantly have to defend themselves from four accusations: of being racist, of being imperialist, of masking social inequalities and of making catastrophic predictions which never come to pass. These accusations are sometimes unfair. But one can pardon the accusers if the source of their allegations is the 1967 bestseller by the Paddock brothers. Agronomist William and US diplomat Paul combined their efforts in this short book, which would be a source of inspiration for the biologist Paul Ehrlich and the climatologist Stephen Schneider. Its title? *Famine, 1975! America's Decision: Who Will Survive?*[12]

Having built up a dataset, the Paddock brothers predicted that the following years would see peak growth in global population and a drastic fall in agricultural production. They argued that global famine was inevitable by 1975 at the latest, but probably before that. Cuba, China and the USSR were, they claimed, already on the brink of collapse. Europe wasn't in good shape either, not to mention India or Africa. The recently decolonised peoples would prove incapable of living democratically (as their 'infantile' anticolonial revolts supposedly demonstrated), and their states would reach the brink of collapse.[13]

In short, there was just one country capable of feeding its population and producing an agricultural surplus: the United States of

11   George Bush, cited in Robertson, *Malthusian Moment*, p. 164.
12   William Paddock and Paul Paddock, *Famine, 1975! America's Decision: Who Will Survive?*, Boston: Little, Brown, 1967.
13   Ibid., p. 107, 108, 110, passim.

America. The existential choice thus posed (and the Paddock brothers fancied themselves as advisors to the imperial prince) was the matter of who would be kept alive and who would be left to die. As against those tender-hearted dreamers who would use the US agricultural surplus as humanitarian aid,[14] the Paddock brothers posed as realists: it was necessary to make a choice. To this end, they enlisted military medicine and its practice of triage. On the battlefield, it would be criminal to try to save those who were already lost, or to try and limit their suffering. The responsible doctor must, instead, concentrate his interventions on those whom he still stood some chance of keeping alive.

This martial metaphor was applied to the nations of the world, with Washington as the medical corps.[15] The Paddocks then offered a review of which countries the US ought to help and which it should abandon. Was it much of a surprise that the first case they analysed was Haiti? This country is the historic symbol of emancipation, of the capacity of men and women of colour to free themselves from the slaveowner's 'wardship' and to govern themselves.[16] The Paddocks' imperial outlook insisted that Haiti was a failure:

> There is nothing whatever in sight that can lift up the nation, that can alter the course of anarchy already in force for a century . . . Now it is too late for an energetic, nationwide birth control program . . . The people are sunk in ignorance and indifference, and the government is entrapped in the tradition of violence. [17]

---

14   Of course, it was nowhere mentioned that exports of US agricultural surpluses above all served the interests of heavily subsidised US producers and destroyed the agriculture of developing countries through dumping.

15   Paddock and Paddock, *Famine, 1975!*, p. 207.

16   On the fundamental importance of the Haitian emancipation struggles, see C.L.R. James's classic *The Black Jacobins: Toussaint Louverture and the San Domingo Revolution*, New York: Dial Press, 1938; and, on its impact on Western thought, Susan Buck-Morss, *Hegel, Haiti, and Universal History*, Pittsburgh: University of Pittsburgh Press, 2009.

17   Paddock and Paddock, *Famine, 1975!*, p. 213.

Their judgement on India (a country whose government was 'among the most childish and inefficient'), Egypt (a country with 'no tradition within the people of being able to govern themselves and to solve their own problems – and the blame cannot be laid on the Turkish and British colonialisms') and a number of other countries in the Global South was of the same paternalist and racist stamp.[18] US help would have to go to its allies, especially those that did not give in to communist blackmail, and to countries that could be expected to control their populations. The global food crisis was put at the service of the US imperial project: 'out of the experience of the Time of Famines' would come 'the foundation on which man built an era of greatness, an era of greatness not for the United States alone but also for the hungry nations.'[19]

Despite its wilder excesses, over the years following its publication the Paddock brothers' book continued to be very widely used by neo-Malthusians. When it came to understanding the impacts that climatic fluctuations could have on agricultural production and thus on living conditions, their analyses would bear real influence.

COLD WAR AND CRISIS OF CAPITALISM

The convergence of environmental, security and population questions in the US in the post-war decades was thus an expression – and in part a critique – of that country's imperial project.[20] But if these responses looked at things through an American lens, what was really at stake was the survival of the capitalist system itself. This problem agitated a coterie of captains of industry preoccupied with the possibility of maintaining long-term economic growth, even under the constraints imposed by the exhaustion of natural resources. The Club of Rome charged Jay Forrester, a veteran of automated defence systems in the United States, with using his military know-how to model the

18   Ibid., pp. 213, 217–18.
19   Ibid., p. 248.
20   Linner, *The Return of Malthus.*

global environment.[21] Like most pioneers in simulating environmental systems, Forrester had been trained in cybernetics, decision sciences, game theory and systems theory. These sciences were supposed to make it possible to predict and understand the USSR's behaviour in the global geopolitical game. Forrester, backed up by young professor Dennis Meadows's team at MIT, used this system to predict the global evolution of resource consumption. These MIT studies mark the point at which the Cold War met with environmental catastrophism. The models used for resource consumption were similar to those the RAND Corporation had used to model catastrophic military scenarios.[22]

The Limits to Growth, published in 1972, was the second study by this research group, following a work rhythm designed to influence the debates of the first UN conference on the environment, held in Stockholm that same year.[23] This book had a planet-wide impact, and it continues to influence the terms of debate even today. Concern for the good long-term management of natural resources is necessary for capitalism. The vast post-1945 development of the state's capacities for planning and intervention in resource management vividly illustrates what a mistake it is to take market apologists at their word and confuse capitalism with the free market. As Polanyi showed, state intervention was essential to establishing a laissez-faire economy; and a market society would be impossible without it. The state has always been the privileged, necessary site of the capitalist class's action to extend its power and create the conditions for capitalist reproduction, as it seeks to extend both horizontally (more sectors of the economy, more territories, etc.) and vertically (within work, within domestic life, in leisure time, etc.).[24] Since the end of the Second World War, the

21   On Forrester and the Club of Rome, see Hamblin, *Arming Mother Nature*, especially chapter 7.

22   Hamblin, *Arming Mother Nature*, pp. 175–6.

23   Donella H. Meadows, Dennis L. Meadows and Jorgen Randers, *The Limits to Growth: A Report for the Club of Rome's Project on the Predicament of Mankind*, New York: Universe Books, 1973.

24   Ralph Miliband, *The State in Capitalist Society*, London: Weidenfeld and Nicholson, 1969.

US state has not only played this role domestically, but, as an imperial state, it has sought to guarantee the conditions for the reproduction of capitalism around the world.[25]

In the late 1960s and early 1970s, the US government, along with its close European allies, pressured for the treatment of environmental questions to become 'global'. This strategy set them apart from the developing countries and the USSR, who instead preferred to defend countries' sovereignty over their own resources. Under the direction of the Canadian industrialist Maurice Strong (who made his fortune in oil), the 1972 Stockholm conference and the subsequent creation of the United Nations Environment Programme (UNEP) marked this global turn in environmental management.[26]

The attempts by the Club of Rome, Maurice Strong and other 'enlightened reactionaries' to reflect on the physical limits on growth never broke with the capitalist logic. The will even to create a global governance of the economy – an overt and external (rather than latent and internal) capitalist planning – was not alien to the debates of the 1970s, though for some economists such an initiative was already 'socialist' in its methods. Nonetheless, the logic of the Club of Rome report, like that of other catastrophist predictions in these years, was to suggest measures to preserve resources and reduce the population in order to save capitalism from itself. At their conference in Philadelphia in 1976 the representatives of the Club of Rome went so far as to declare:

> In the conditions of the coming times, the necessity of re-establishing a viable world balance between population and resources – if the present generation is not able to timely adopt the necessary corrective measures – will inevitably tend to bring about a technocratic version of oriental despotism, of which Stalinism and Nazism have already given us an anticipated view.[27]

---

25 See Panitch and Gindin's important *The Making of Global Capitalism*.

26 Hamblin, *Arming Mother Nature*, pp. 194–6.

27 *New York Times*, 19 April 1976, cited by Stephen William Rousseas, *Capitalism and Catastrophe: A Critical Appraisal of the Limits to Capitalism*, Cambridge: Cambridge University Press, 1979, p. 93; author's emphasis.

As the economist Stephen Rousseas comments, this choice between reformed capitalism and 'oriental despotism' strangely resembles (albeit in inverted fashion) the alternative which Rosa Luxemburg presented between socialism or barbarism. For the likes of Kissinger or Brzezinski, the adoption of an openly authoritarian regime, in order to ensure the continued supply of natural resources in an era of growing scarcity, was a real possibility.

THE VULNERABILITY AND DECLINE OF CIVILISATION

The fear of looming ecological collapse extended beyond neo-Malthusian theses, although it was linked to them. The US defeat in Vietnam as well as the threat of nuclear war bolstered intense feelings of vulnerability and, in turn, the determination to create bulwarks of protection against shocks. To conservatives' eyes, the state seemed less and less capable of guaranteeing the security and reproduction of the social order. For liberals, consumer society and the urban way of life brought new risks: individuals' growing attachment to complex systems would render them more vulnerable. The city dweller depended on the electricity network for heating, the transport network to get around, the water systems to drink, and so on. For instance, the OPEC countries' embargo in 1973–74 revealed the US dependence on oil imports.

In an article published in the *New York Times* in 1977, Amory Lovins denounced the vulnerability of the US electric grid, especially in the region of New York.[28] The very size of these networks, and their complexity, left them ungovernable and liable to collapse. Centralised energy production put the population at the mercy of an enemy, who would only need to control the centre of this production in order to set off a major crisis. Lovins worried over the British miners' strike and the capacity that a union which controlled the electricity grid would have to bring the country 'to its knees'. He suggested that the response to this vulnerability lay in developing the decentralised

---

28   Amory Lovins, 'Resilience in Energy Strategy', *New York Times*, 24 July 1977.

production of renewable energy, which would increase the resilience of the energy system and its capacity to cope with shocks.[29]

As well as this diffuse array of civilisational threats, vulnerabilities, crises and expressions of decline, an extra reason for worry came from the now-rising climate sciences. There was serious discussion of the possibility that certain enemy countries, starting with the USSR, would seek to change the climate deliberately and thus make it into a weapon of war.[30] But most important, the unintentional yet catastrophic changing of the climate by humans added to the feeling of collapse which many scientists and political analysts now experienced. Specialists in the atmospheric sciences became particularly prone to make the climate and its changes into an all-powerful element which would decide the future of human civilisations.[31]

Popular science books on the climate threat, which flirt with science fiction, multiplied, as did interventions in more serious media.[32] According to the philosopher Peter Timmerman, some of the climate debates in this period bore more than a whiff of Spengler's theses on the decline of civilisation.[33]

Yet until the early 1980s, the thesis of global warming due to humans' greenhouse gas emissions had not yet fully established itself. Scientists who dealt in climate questions spoke above all of growing climatic variability. If most began to share the analysis that there was

29   As Timothy Mitchell has shown in *Carbon Democracy*, trade-union struggles around energy production were one of the great motors of democratisation in the twentieth century. But they also prompted strategies on the part of businesses and states seeking to quash workers' power at the point of production. Lovins's anti-union analysis is part of this tradition, though Mitchell makes no reference to this.

30   See Hamblin, *Arming Mother Nature*, and Fleming, *Fixing the Sky*.

31   This claim was reported with some amusement by the participants in the SCOPE programme seminar on climate/social relations held in Toronto in 1978. See the report in ICSU/SCOPE, *ICSU/SCOPE Workshop on Climate/Society Interface, December 10–14, 1978, Toronto*, Paris: SCOPE Secretariat, 1978, p. 9.

32   To such a point that even the climatologist Stephen Schneider, himself hardly reticent about indulging in climate catastrophism, was troubled by them: see Stephen Schneider, 'Against Instant Books', *Science*, vol. 270, no. 22/29, 1977, p. 650.

33   Peter Timmerman, *Vulnerability, Resilience, and the Collapse of Society: A Review of Models and Possible Climatic Applications*, Toronto: Institute for Environmental Studies, University of Toronto, 1981, and personal communication, 20 May 2015, Toronto.

global warming, a minority saw the possibility of global cooling. In any case, in this era not all scientists accepted the existence of a link between fossil fuels and global warming.

The consequence of these uncertainties was that, despite the catastrophist forecasts, few analysts in this era asserted the need to get a handle on climate change or 'control' it by reducing greenhouse gas emissions. On the contrary, most took climate change for granted and reflected on the ways in which societies would suffer its impacts or manage to adapt to it. The early political debates on climate change hardly ever sought to attenuate or reduce greenhouse gas emissions, but rather more to anticipate their impact and learn to live with these changes – or even to profit from them. As Joshua Howe aptly writes, 'Theirs was a gospel of preparedness.'[34]

CAPITALISM'S ADAPTATION TO THE CLIMATE

'Is there hope for man?' Such was the dismal question that began the left-wing economist Robert Heilbroner's book *An Inquiry into the Human Prospect*.[35] Great philosophical treatises on man or the human condition strike our relativist tastes as outdated. After the postmodern critique of the 1980s and 1990s, it seems incongruous to claim to speak in the name of a single humanity sharing common characteristics. But Heilbroner was writing in 1974, at the end of three decades in which US intellectuals had continually been serving up dissertations on the crisis of humanity and civilisation. Hannah Arendt and her *The Human Condition* (1958) Lewis Mumford and his *The Transformations of Man* (1956) and Herbert Marcuse and his *One-Dimensional Man* (1964) are the most famous testaments to this type of literature, abundant in this period.[36]

---

34   Howe, *Behind the Curve*, p. 115.

35   Robert Heilbroner, *An Inquiry into the Human Prospect*, New York: W. W. Norton and Co., 1974.

36   Christopher Benfey, 'The Case of the Skeptical Pragmatist', *New York Review of Books*, vol. 62, no. 11, June 2015, pp. 53–6, citing Mark Greif, *The Age of the Crisis of Man: Thought and Fiction in America, 1933–1973*, Princeton, NJ: Princeton University Press, 2015.

While Heilbroner did not enjoy the posthumous aura of his friend and mentor John Kenneth Galbraith, he was a front-rank public intellectual. Author of a famous textbook providing an introduction to economics, a commentator in the big dailies and a verbose writer, his book was received with interest and was a publishing success, despite his pompous style.[37] Borrowing from the neo-Malthusians, Heilbroner pointed to the constraints which presented themselves to all governments in a world characterised by finite resources. He prophesied a new authoritarianism and a growing intrusion of state regulations into everyday life in the name of the environment.

If reading this *Inquiry* is not a waste of time, it is because amid its rather conventional reflections on the end of progress, the halting of economic growth, nuclear war and human overpopulation – and the book says nothing original on these points – Heilbroner identifies a major problem for humanity: global warming. He was then counting on the existence of a 150-year interval before this warming began to make itself felt; but in his eyes, only immediate action now could stop the growth which was responsible for the coming disaster.[38]

Heilbroner used climate catastrophism to shed a different light on what was a recurrent debate in this era: the relative survival prospects of capitalism and socialism (meaning the Soviet bloc). He noted that 'the issue of the relative resilience and adaptive capacities of the two great socio-economic systems [had] com[e] to the fore as the decisive question'[39] How did global warming make it possible to judge these two opposed systems' adaptive properties?

Faced with Richard Falk's line of questioning on capitalism's ability to adapt to the ecological crisis, Heilbroner provided a pessimistic response. It was the communist states, he thought, which had the

37   James Fallows, 'How Do You Protect the Environment Without Interfering with Privacy?', and Wade Greene, 'Economists in Recession: After an Inflation of Errors and a Depletion of Theory', *New York Times*, 12 May 1974.

38   Heilbroner, *Human Prospect*, pp. 50–5. We should nonetheless note that the climate warming mechanism identified by Heilbroner is not the concentration of greenhouse gases in the atmosphere but a rise in temperature owing to human activities (heating, burning fuel, etc.).

39   Ibid., p. 130.

greater capacity to react and immediately adapt. Since capitalism depended on growth, it would not be able to organise a transition to an economy which consumed less energy and fewer materials: 'It appears logical to conclude that socialism, with its direct commitment to a planned economy and with its freedom from the ideological blockages of private property, could manage the adaptation of an industrial society to a stationary equilibrium much more readily than capitalism.'[40]

But Heilbroner added that this could only play to socialism's advantage in the short term – and that in a more distant future, Malthusian constraints would also hurt the socialist societies. He expressed his pessimism about the capacity of industrial societies, be they capitalist or socialist, to reform and avoid ecological collapse. Like other New Age catastrophists, he saw salvation only in a post-industrial society, returning to so-called 'primitive' modes of production, abandoning the cities and finding fulfilment in inner reflection and meditation.

Unlike Heilbroner, many economists argue that capitalism's superiority in fact lies precisely in its greater ability to adapt to changing circumstances. Two schools of thought seem to clash on this question. The catastrophist school, examined in this chapter, generally doubts the market's capacity to resist climatic shocks. The other school, addressed in chapter 2, instead looks to the market for the flexibility needed to adapt capitalism to new constraints (as opposed to the bureaucratic lumbering attributed to the Soviet system).[41]

CLIMATE IRREGULARITIES AND FAMINE IN THE SAHEL

The prolonged drought in the Sahel from 1968 and 1973, and the deadly famine which followed, seemed to provide brutal confirmation of neo-Malthusian theses on the collapse of agricultural production. Further climatic imbalances in this same moment had a

---

40   Ibid., p. 90.
41   Hamblin, *Arming Mother Nature*, p. 178.

brutal effect on rural populations, especially in India. Climatic variations again seemed to have a decisive impact on food production, also bringing a series of afflictions from famine to social turmoil, forced migration and so on.[42] But, for two decades, the episode in the Sahel would above all haunt the thinking of climatologists and ecologists. All popularising works on the climate, and a good share of scholarly contributions, would invariably begin with the episode in the Sahel, as famines began to be attributed to fluctuations in the climate.[43]

The convergence between neo-Malthusian theories on the booming population, the exhaustion of food production and climatic change became increasingly clear at the international conferences held under the auspices of the United Nations across the 1970s: the Stockholm conference on the human environment in 1972, the global conference on food in Rome in 1974, the UN conference on water in Mar del Plata in 1977, and a further one on desertification held in Nairobi that same year.[44] The 1974 conference on food had its origins as a response to an appeal by the Algerian president Houari Boumédiène. The countries of the Global South hoped that there they would be able to debate the structural causes of malnutrition and poverty, the harmful role played by multinational companies, and the worsening terms of trade. US Secretary of State Henry Kissinger skilfully managed to avoid the subject, to the benefit of the agri-food industry: the only discussion was of urgent measures to take in case of famine, not of the causes of malnutrition.[45]

---

42 Howe, *Behind the Curve*, p. 103.

43 For instance, John Gribbin (ed.), *Climatic Change*, Cambridge: Cambridge University Press, 1978; and Margaret Biswas and Asit K. Biswas (eds), *Food, Climate, and Man*, New York: Wiley, 1979. See also the works by Schneider and Bryson discussed below. We will go on to discuss the critique which radical geographers made of analyses of these famines, holding them to be essentially natural or climatic catastrophes.

44 On this influence, see F. Kenneth Hare, 'Climate: The Neglected Factor?', *International Journal*, vol. 36, no. 2, 1981, pp. 371–87.

45 Erich Jacoby, 'En finir avec les "fléaux de la pauvreté de la malnutrition et de la faim?"': Après la conférence mondiale de l'alimentation', *Le Monde diplomatique*, December 1974.

The delegates to the Rome conference were given a copy of Lester Brown and Erik Eckholm's book *By Bread Alone*.[46] Brown was an agronomist who had in 1974 founded, with Rockefeller Foundation support, what would become one of the main environmental think-tanks used by the US government and international organisations, the Worldwatch Institute. In *By Bread Alone*, Brown and Eckholm emphasised overpopulation as well as the climate-related causes of food poverty, especially the southward movement of the Sahara. They attributed this phenomenon of desertification to the overexploitation of the soil by poor peasants. In the same vein as the Paddock brothers, they warned against the illusion that adequate food reserves would be available in perpetuity. Most important, drawing on the works of the climatologist Reid Bryson,[47] Brown and Eckholm advanced, first, the idea that human activities were now acting on the climate; and, second, the idea that climatic fluctuations were increasing and would threaten international trade given their effect on food production. As well as a reduction in the global population, they recommended (as would Schneider) the creation of world grain reserves in order to adapt to possible variations in the climate.[48] As at the end of the nineteenth century, the climate question thus became an explanation of famine and poverty,[49] while at the same time serving as an argument for the creation of institutions that would make it possible to reduce populations' food vulnerability.

---

46   Lester R. Brown with Erik P. Eckholm, *By Bread Alone*, New York: Praeger, 1974.

47   Reid A. Bryson, 'A Perspective on Climatic Change', *Science*, vol. 184, 1974, pp. 753–60.

48   A viewpoint also defended by the climatologist Walter Orr Roberts, but criticised by the agronomist René Dumont and the economist Ignacy Sachs, for whom the grain reserves controlled by the United States are but a further means of increasing the dependence of Third World countries and strengthening imperialism. For these different points of view, see Aziz Sartaj (ed.), *Hunger, Politics and the Markets: The Real Issues in the Food Crisis*, New York: New York University Press, 1975.

49   Mike Davis, *Late Victorian Holocausts: El Niño Famines and the Making of the Third World*, London and New York: Verso, 2000.

VARIATIONS IN THE CLIMATE AND
GEOPOLITICAL DESTABILISATION

From 4 to 8 June 1975, the Rockefeller Foundation organised a symposium at its conference centre in Bellagio, on the edge of Lake Como, studying the links between climate change, food shortages and international conflicts. Kenneth Hare invited along the climatologists Walter Orr Roberts, founder of the National Center for Atmospheric Research (NCAR) and Reid Bryson, as well as the UNEP's deputy executive director Mostafa Tolba. Among these experts, the hypothesis of human-caused global warming had become central for explaining growing climatic variability, though they examined this on an equal footing with other theories, in a context in which the possibility of global cooling was also entertained. The essential idea was that fluctuations in the climate induced fluctuations in global agricultural production and thus famine risks. There was talk of absolute grain shortages in certain countries. This could provoke conflicts between states as hungry countries sought to secure their food supply from others which were in surplus.[50] Hare captured the mood of the conference as he stated, 'Weather has re-emerged, in fact, as a major destabilizer of the world economic system'.[51]

The following year, the US National Academy of Sciences would produce its own report on the question, entitled 'Climate and Food'. But despite these various activities, climatologists thought that agronomists and economists remained overly reluctant to accept that climate change was a decisive factor for agricultural production. Kenneth Hare regretted the fact that the task of drawing the public's attention to the agricultural consequences of global warming had fallen to a climatologist, Stephen Schneider, whose *The Genesis*

---

50   *Climate Change, Food Production, and Interstate Conflict: A Conference Held at the Bellagio Study and Conference Center, Italy, 4–8 June 1975*, Working Papers, New York: Rockefeller Foundation, 1976.

51   F. Kenneth Hare, *Climate Change, Food Production, and Interstate Conflict*, p. ix.

Strategy: Climate and Global Survival published in 1976,[52] became a bestseller.[53]

Reid Bryson was, along with Schneider, one of the main popularisers of climatologists' research in the 1970s.[54] Both emphasised that the climatic variations of the late 1960s and early 1970s were not 'normal' but signalled a more important change in the global climatic system. Most importantly, they were eager to demonstrate its social and political consequences. Bryson published a book together with his colleague Thomas Murray (with a preface by Hare) explaining the evolution of human history on the basis of variations in the climate.[55] Writing in a neo-Malthusian vein, they explained famines in terms of the carelessness of the poorest, who over-farmed the soil and thus ruined it. They added that the climate, alongside overpopulation, was a major factor for food insecurity.

But Bryson and Murray did not believe that the natural and social factors that altered the climate could be controlled: 'We can't very well stop Third World people from farming their land and polluting the air, and we probably can't or won't stop the activity in our industrialized lands. Perhaps, though, we can begin to realize the limitations that the earth, with its sequence of climates, places on our numbers and our actions'.[56]

If it was impossible to do anything about the causes of climate change, the implication was that the only reasonable attitude was to adapt behaviour to its consequences. For dogged neo-Malthusians like Bryson and Murry, this naturally meant reducing the human population.

---

52 Stephen H. Schneider and Lynne E. Mesirow, *The Genesis Strategy: Climate and Global Survival*, New York: Plenum Press, 1976.

53 Hare, *Climate: The Neglected Factor?*, p. 377.

54 For instance, Weart, *Discovery of Global Warming*, pp. 86–90, passim.

55 Reid A. Bryson and Thomas J. Murray, *Climates of Hunger: Mankind and the World's Changing Weather*, Madison: University of Wisconsin Press, 1977. The authors thanked Stephen Schneider for reading the manuscript before publication.

56 Ibid., p. 156.

CLIMATOLOGY, A MISSIONARY FOR THE AGRIFOOD INDUSTRY

Neo-Malthusian analysis could, nonetheless, also lead to diametrically opposite conclusions. 'World hunger' – as well as the USA's saviour mission – were at the heart of the gigantic transformation of Third World agriculture to which the Ford and Rockefeller Foundations devoted themselves from the 1950s onward, acting on the mandate of the US government. This 'green revolution', which Lise Cornilleau and Pierre-Benoît Joly define as 'a US programme for spreading Western agricultural technologies'[57] also rested on the idea of a shortage in global agricultural production. But, unlike the crude neo-Malthusians, the partisans of the green revolution thought that technological developments would make it possible to break out of Malthusian scarcity by raising agricultural productivity. To that end, these foundations encouraged the mechanisation of agriculture, the choice of high-yield crops and mechanised export agriculture with multiple chemical fertilisers. In short, the green revolution sought the abandonment of peasant agriculture and the development of an industrial, capitalist agriculture. As against the socialist intentions of many of the Third World governments that had just achieved independence, this 'revolution' would make the Southern countries' agricultural production dependent – and enduringly so – on the multinationals that controlled the global markets, and especially American firms.

In this context, the argument for adaptation to climate change was running at full tilt. The astronomer and physicist Walter Orr Roberts, founder of the US NCAR, insisted that the climate's impact on food production and food security required better study, and that technological solutions could indeed be developed. He advanced his views in a 1973 seminar on 'climatic limits' organised in Aspen, Colorado, by UNEP executive director Maurice Strong, following the

---

57   Lise Cornilleau and Pierre-Benoît Joly, 'La Révolution verte, un instrument de gouvernement de la "faim dans le monde": Une histoire de la recherche agricole internationale', in Pestre (ed.), *Gouvernement des technosciences*, p. 172.

Stockholm conference. Amory Lovins was the rapporteur at this conference, chaired by the Canadian climatologist Kenneth Hare. Roberts managed to persuade the Rockefeller Foundation (Strong was one of its trustees) to finance research in this field. From 1974 to 1981, he directed the Aspen Institute's programme on 'food, climate and the world's future'.[58]

If climatologists and agronomists from the US and Canada were at the spearhead of this research, backed by big private foundations and public institutes, this should be explained in terms of North America's central position in global food production. The effects of climatic variations on food production struck right at the heart of Midwestern farmers' interests, as well as those of the US and Canadian agrifood industry – an industry able to establish neocolonial control over global food markets.

In his role as head of the Aspen Institute programme, Roberts made climatology a missionary science in service of US agrifood companies' interests. As he put it:

> The Program seeks to strengthen the role of industry in providing answers to some of the difficult world food issues that confront us as a consequence of global climate fluctuations ... The Program is designed to look cooperatively with the participating companies at the major long range planning and policy issues facing the companies, the nation and the World.[59]

Thus, as distinct from more fundamental research programmes on the climate system, Walter Orr Roberts vaunted his project's practical approach. While based on cutting-edge advances in climatology – and he also emphasised his own experience in this regard – his project above all sought to provide practical knowledge to stakeholder companies, so that they could bring to bear appropriate influence on

---

58   Hare, *Climate: The Neglected Factor?*, pp. 374–6.

59   Walter Orr Roberts, 'A Food and Climate Program. In Cooperation with a Group of American Corporations', Memo, 12 July 1977, p. 1, AAAS, box 21.

national and global food and agricultural policies. The programme was supposed to bring together the best technical experts and representatives of stakeholder companies each year for the purposes of debate and exchange. For instance, it promoted discussions of the constitution of global food reserves, the role that the US should play on account of its 'agripower', the development of plants adapted to fluctuations in the climate, etc.

But beyond immediately practical questions, what also stuck out were certain existential questions for these firms: 'Is the free enterprise system viable in the long term if a large segment (minimally twenty-five percent) of the world's population continues to experience malnutrition?',[60] Roberts asked. For him, the question of the growth in the global population was central. Moreover, he allowed his neo-Malthusian orientation to seep through when he listed important questions that ought to be addressed, such as 'a. Triage? (Let the poorest third die in order to save the middle third.) b. Food only in return for population control measures?'[61]

A dozen companies financed the project. These included the farm vehicle and equipment manufacturer Deere & Company, the brewer Adolph Coors, the fertiliser producer CF Industries, the agrifoods conglomerates Mars Inc. and Norton Simon Inc., the cigarette firm Philip Morris Inc., and so on. In Canada, another farm equipment manufacturer, Massey-Ferguson, launched its own research and communications programme in this same period, addressing 'climate variability and its relation to food production'. But the Aspen programme sought above all to serve as an outrider for the interests of agrifood companies in Washington, especially in extending their businesses into the Third World.

For instance, in August 1978 the Aspen programme organised a week-long workshop on 'private/public collaboration and Third-World food systems'. Financed by US development agency USAID, it brought together not only specialists working on agronomy and

---

60   Ibid., p. 12.
61   Ibid., p. 6.

development, but around 40 industry representatives who wanted
their government to provide more emphatic support for US multinationals seeking to expand into Third World countries.[62] This
demand was justified as a contribution to the fight against malnutrition in the lands of the Global South, in a context of overpopulation
and underinvestment. The participants seemed especially worried by
'Third World attacks on transnational companies'. They called on the
US State Department to enact a series of twenty recommendations
which were supposed to reduce the risks to US investors who sought
to penetrate these countries' food markets. They suggested, for example (recommendation no. 7): 'The U.S. government should encourage
more participation by the private sector in the problem of nutritional
needs and consumer education in LDCs [least developed countries]
– particularly involving women'; or (no. 11): 'The U.S. government
should develop a clearly stated policy to favor and encourage participation of U.S. agribusiness in planning and implementing development programs in LDCs.'[63]

That same year, Roberts gave the programme's 'distinguished
lecture' on the theme of mankind's impact on the climate. He again
emphasised that not only was climate variability increasing, but that
global warming should be expected, with dramatic consequences for
global agricultural production. He concluded that corporations, aided
by the Aspen Institute programme, should be prepared 'to cope with
these changes in our social, political, economic and ethical systems
that will result from climate changes of this magnitude.'[64] What was
now called for, more than ever before, was a proactive adaptation to
the changes underway. And those who were able to anticipate it would
come out better than the rest.

---

62   Mary L. Wolff and Lloyd E. Slater, 'Report of the Workshop on Private/ Public
Collaboration and Third World Food Systems; 20–26 August 1978, Aspen Colorado',
Program on Food, Climate and the World's Future, Aspen Institute for Humanistic Studies,
Boulder, 5 September 1978, AAAS Archives, box 21.

63   Ibid., pp. 13, 17.

64   Walter Orr Roberts, 'An Inquiry into Man's Impact on Climate', The Food and
Climate Forum Distinguished Lecture 1978, Aspen Institute for Humanistic Studies,
Boulder, AAAS Archives, box 21.

THE GENESIS STRATEGY

Steve Schneider was one of the first climatologists to play the card of public opinion. Unlike most of his colleagues, who were happy to publish in specialist journals and keep to their own narrow domain of expertise, Schneider sought to share the discoveries of the atmospheric sciences among a wider audience. Above all, he sought to contribute to the debates on the global 'problem' to which Heilbroner, the Paddock brothers and the Club of Rome had already pointed.

Schneider's book *The Genesis Strategy: Climate and Global Survival*, published in 1976, mixed a striking quantity of information with political proposals for the way out of this situation. In addition to the existing discussions on global population, food production and environmental degradation, Schneider addressed the question of the climate. His book dealt with the pollution of the atmosphere as well as damage to the ozone layer. But the young climatologist concentrated his attention on human-caused climate change. He noted the importance of fossil fuel-caused CO_2 emissions for changes in the atmosphere. Nonetheless, Schneider still did not explicitly identify with the thesis of global climate warming, even though he did briefly reference this idea.[65] He instead lay greater emphasis on changes in the climate – and in particular, the rise in global climate variability.

Unlike the more conservative neo-Malthusians like the Paddock brothers or Garret Hardin, Schneider did not cast overpopulation as the only variable. His political affinities were 'liberal', in the US sense of the word. *The Genesis Strategy* was peppered with jibes against US militarism, against the power of major corporations and – following in Polanyi's footsteps – against the belief that the 'free' market would succeed in securing the food supply. He instead foregrounded the precaution principle and citizen participation in the deployment of technoscience. He highlighted the problem of global development inequalities and the need for compensation for people in the Global South who were suffering the consequences of climatic transformations.

---

65 Schneider, *Genesis Strategy*, p. 179–82.

Yet in this 1976 text, the idea that the global atmospheric changes now underway could be held back by policies to cut greenhouse gas emissions was notable by its absence. Only the climate at a local or regional level could be altered by intentional means. The human and natural impacts on the global climate were taken for granted, as were their effects: famine, displaced populations, conflicts, etc. Hence the only possible strategy was the one evoked in his title: a strategy of preparing for catastrophe. By 'Genesis' he referred to the episode in the Bible where the prophet Joseph announces to the Pharaoh that seven years of plentiful harvests would be followed by seven years of dearth. As a result, the Pharaoh ordered Joseph to organise food reserves, in order to allow the population to get through the times of hardship.

Similarly, Schneider acknowledged the need to build up global food reserves in order to deal with the famines which were now coming on account of climate variability. He wrote, 'proper planning for controlled growth is our best alternative, at least for the long term.'[66] At a deeper level, his Genesis strategy was a strategy of resilience, which demanded a diversification of food and energy sources in order to avoid dependence on any single production system. For Schneider, the main problem was the need for political action to deal with uncertainty – an uncertainty which climate variability would itself massively intensify.

But the climatologist also formulated solutions for the global 'problématique' (as he wrote, in French). Criticising the democracies' short-termism as well as the perverse incentives associated with politicians' quest for re-election, he suggested various mechanisms for government by experts. The main one was the creation of a 'fourth political power', a branch of government responsible for 'truth and consistency', to be made up of scholars who would sift through the different political options in order to determine their future consequences. They would be aided by various committees such as an 'Institute for Imminent Disasters', an 'Institute for Available Resources'

66   Ibid., p. 241.

and an 'Institute for Political Options'.[67] Nonetheless, for Schneider, only a global intergovernmental agreement backed up by global, expert-staffed planning institutions (agronomists, climatologists, demographers, political scientists, etc.) would be able to re-establish a global social, economic and climate equilibrium.[68]

Schneider's book was lauded by the main popular scientists of the era, like Paul Ehrlich, Carl Sagan and Lester Brown (who were also friends of Schneider's). But the media success of this work – he was repeatedly invited onto television's *Tonight Show* and saw his book receive full-page coverage in both the *New York Times* and the *Washington Post* – discomfited some of his climatologist colleagues.[69] Yet even this relative coolness did not have negative consequences for Schneider's career. His expertise and institutional influence would continue to be a major force in climate science circles, especially in the IPCC, up till the 2000s.

Shortly after the publication of *The Genesis Strategy*, Schneider revisited the question of the fossil fuel–induced build-up of CO_2 in the atmosphere. This time around, he had no hesitation in speaking of global climate warming. With Robert Chen, he detailed its consequences in terms of rising sea levels and the threats to coastal areas, especially in Florida.[70] From this point onward, he presented the political options in the following terms. Adaptation is a passive solution: it is the spontaneous result of the actions human societies take in response to climate change.[71] In counterposition to this, he proposed 'building resilience' as a proactive and deliberate form of response,

67   Ibid., pp. 299–311.

68   Ibid., p. 303.

69   Howe, *Behind the Curve*, pp. 100–3.

70   Stephen H. Schneider and Robert S. Chen, 'Carbon Dioxide Warming and Coastline Flooding: Physical Factors and Climatic Impacts', *Annual Review of Energy*, vol. 5, 1980, pp. 107–40.

71   This is also the opinion, among others, of the philosopher Klaus M. Meyer-Abich, 'Socioeconomic Impacts of CO_2-Induced Climatic Changes and the Comparative Chances of Alternative Political Responses: Prevention, Compensation, and Adaptation', *Climatic Change*, vol. 2, 1980, and of the political scientist Michael Glantz, 'A Political View of CO_2,' *Nature*, vol. 280, 1979, pp. 189–90.

seeking to minimise the impacts of global warming, i.e. pursuing the Genesis strategy.

Lastly, he considered the possibility of reducing CO_2 emissions: in increasing energy efficiency, in developing the use of renewable energy, in mastering territorial management, and so on. But Schneider was sceptical about the likely success of such a choice, for CO_2 emissions were the inevitable other side of the coin of development. Further, he remarked that the costs of reducing emissions would have to be paid immediately, whereas the costs resulting from the damage caused by global warming would only later make themselves felt. He thus concluded that the present generations were not ready to pay too high a price for reducing emissions, relative to benefits that lay only far in the future. Conversely, measures taken to build up resilience, or out of proactive adaptation to global warming, would bring immediate, tangible benefits also for the present generations:

> Even if there were no CO_2 problem, the activities called 'build resilience' are important for other reasons. Viewed this way, the CO_2 issue has tie-ins with other problems. In view of the large CO_2 uncertainties, it seems likely that policy responses with tie-ins are the ones that will receive priority attention by decision makers[72]

In other words, it was far more likely that the response to the problem of climate change would revolve around proactive adaptation than reducing carbon emissions.

BOGGED DOWN IN OIL

Climate imbalances would also affect another aspect of US politics – energy policy. The oil shocks of the early 1970s convinced President Richard Nixon of the need to guarantee US supply sources in the long term. He encouraged the expansion of resource supplies

---

72   Schneider and Chen, 'Carbon Dioxide Warming', p. 134.

from within US territory by stepping up nuclear production. He also backed energy saving and renewable energy (solar, wind, etc.) Following the Arab oil producer countries' embargo in 1973, Nixon launched the 'Independence' project seeking to achieve US self-sufficiency in energy production by 1980. But with his forced departure following the Watergate scandal, it was his successor Gerald Ford who cut the ribbon for the Energy Research and Development Administration (ERDA) in 1974. This public agency soon initiated a series of studies, preliminary to the establishment of natural energy planning.[73]

One of these studies was commissioned from a US research institute, Stanford Research International (SRI), and entrusted to a futurology specialist, Oliver W. Markley. This study concerned the 'Sociopolitical Impacts of a Carbon Dioxide Buildup in the Atmosphere Due to Fossil Fuel Combustion.'[74] The SRI researchers did not settle for summarising the climatology literature. Rather, they also established a grid of socio-political impacts of the rising volumes of carbon dioxide in the atmosphere. Their study predicted growing state intervention in all manner of domains, including social welfare, the regulation of consumption and energy usage. In response to international shortages, they envisaged the development of 'private transnational agribusiness' as well as the possible collapse of marginal nation-states, the introduction of a 'triage' policy for the allocation of global food aid and mounting socio-political conflicts.

The interesting thing about this study shone through from its title: that is, it sought to analyse global warming as a consequence of fossil fuel use. It would be illusory to talk about global warming if one was not also prepared to talk about oil! The analysis of the (catastrophic) effects of global warming would serve as ammunition for

73   Alice Buck, 'A History of the Energy Research and Development Administration', Washington, DC: US Department of Energy, March 1982.

74   Oliver W. Markley et al., 'Sociopolitical Impacts of a Carbon Dioxide Buildup in the Atmosphere Due to Fossil Fuel Combustion', in *Report Prepared for Inexhaustible Energy Resources Planning Study Energy Research and Development Administration*, Washington, DC: Business Intelligence Program, SRI International, 1 July 1977.

those who sought to replace fossil energy with renewables: the greater the effects of warming, the more justified a rapid energy transition would become.[75]

But while the futurologists thus had to recommend a massive reduction in CO_2 emissions, they acknowledged that the weight of the oil industry made such an option impracticable. Moreover, cutting fossil fuel use in a single country – even if that meant the United States – would have almost no effect on the global problem: it would only delay it by a few years. It was thus necessary to adapt to global warming rather than attempt to reduce CO_2 emissions:

> A more gradual cutoff [of fossil power] might be required if CO_2 problems worsened, but the necessary international support for effective CO_2 reduction policies is unlikely to exist. CO_2 increases are more likely to be something with which we must learn to live.[76]

THE CLIMATE: A NEW CRAZE

The US Department of Energy was no longer willing to overlook the climate question. In collaboration with the American Association for the Advancement of Science (AAAS), the leading US scientific body and publisher of the journal *Science*, it financed a research programme on the social and economic impacts of climate change.[77] Roger Revelle and Stephen Schneider were each involved in organising the programme, which in 1979 resulted in the first international conference dedicated to the social sciences of global warming. This seminar, held in Annapolis, Maryland, sought to bring together specialists in economic history, anthropology, economics and political science to think through the consequences of global warming and the responses it demanded: Schneider, Revelle, Kellogg, Orr Roberts, Kenneth Hare and Crispin Tickell all took part. While Emmanuel Le Roy Ladurie

75　Ibid., p. 1.
76　Ibid., p. 51.
77　Howe, *Behind the Curve*, pp. 107–17. See also AAAS Climate Program Records, especially Series I and III, Washington, DC, AAAS Archives.

and Buzz Holling had to decline, economists like Lester Lave, Mancur Olson and William Nordhaus did participate.

The conference addressed a broad array of themes, running from the long history of human societies' adaptation to climate change to the transformation of political institutions in response to the threat posed by CO_2. Almost all the interventions spoke of adaptation, deemed more or less spontaneous. In these debates, there was almost no reference to cutting greenhouse gas emissions (the exception was the jurist Edith Weiss; notably, she proposed the creation of a market in pollution permits).[78]

The natural – and especially climate – sciences provided the framing for the problem; then, in a second phase, the social sciences were called on to imagine what the response should be. The natural world and the social world were sharply separated, even if they did interact in this second phase. The pathetic appeals to interdisciplinarity did little to mask this functional hierarchy of knowledge – one destined to a long future in discussions of the climate question. As the climatologist and historian of science Mike Hulme has demonstrated, this hierarchy in fact typified the determinist vision of climate change which prevailed up until the mid 2000s. It meant expecting the atmospheric sciences to reduce, to the maximum possible degree, any uncertainty in predicting not only the future climate, but also its impacts on society. The increasingly sophisticated modelling of climate change supposes a vision in which society reacts in a highly determinist fashion (for example, through migration, or through conflicts breaking out). But this means ignoring society's capacities for innovation and imagination – and even the indeterminacy of the future itself.[79]

---

78 Other than the archives mentioned, we can see this in the edited volume based on the contributions to this conference: Robert S. Chen, Elise Boulding and Stephen H. Schneider (eds), *Social Science Research and Climate Change: An Interdisciplinary Appraisal*, Dordrecht: D. Reidel, 1983.

79 Mike Hulme, 'Reducing the Future to Climate: A Story of Climate Determinism and Reductionism', *Osiris*, vol. 26, no. 1, 2011, pp. 245–66. Hulme is nonetheless mistaken in claiming (p. 264) that the concepts of vulnerability and resilience were introduced in the late 1990s to early 2000s in order to offer a more socially realistic perspective on the analysis

The following year, 1980, the Department of Energy and the AAAS tied into the Aspen Institute programme on climate, with a seminar at the Oak Ridge atomic research centre. This seminar was meant to propose practical orientations for research into how agriculture in Third World countries could adapt. Under Revelle's direction, these experts from the natural sciences, but also economists (Lave, Nordhaus, Mancur Olson . . .) and sociologists, identified sixteen priority research axes. These axes essentially had to do with agricultural adaptation, the management of water resources, genetic inheritance in plants and the management of food stocks under the constraints imposed by the climate. The worldview which shone through was still a neo-Malthusian one, flecked with culturalist and paternalist assumptions. The final report thus cited the 'typical Third-World villager' whose perception of climate change was troubled by other changes – those brought by the 'technological development of the West' – given his own 'culture', formed by a 'pre-1800' civilisation passed down from father to son. But this villager would have to constantly adapt to a changing environment – and 'adapt culturally'. To help him adapt, the experts in Oak Ridge proposed

that an anthropologist . . . should spend six months with a village family in each of the six major regions of cultural variation: post-Confucian, Malay-Indonesian, Hindu, Islamic, Christian and African tribal . . . suggesting which areas of the Third World will be least equipped to cope with climate change. While its outcome cannot be predicted at this point, it appears likely that the more cohesive cultures, such as the post-Confucianist Far East . . . may be able to adjust to a warmer world much more easily than, say, those societies whose cultures are more deeply religious or otherwise

of climate change. These concepts were already current in the 1970s. For instance, see Timmerman, *Vulnerability, Resilience, and the Collapse of Society*, on the idea of resilience: on vulnerability, see Michael Watts, 'On the Poverty of Theory: Natural Hazards Research in Context', in Kenneth Hewitt (ed.), *Interpretations of Calamity from the Viewpoint of Human Ecology*, Boston: Allen & Unwin, 1983, pp. 231–62; or the works by Rolando García discussed in chapter 2.

traditional (the Muslem [sic] world and tribal Africa being at the other end of the spectrum).[80]

The seminars in Annapolis, Oak Ridge and other similar initiatives showed that applied research in the United States was at the cutting edge not only of climate science, but also in studying the impacts of climate change. Financed by the US administration and the big private foundations, this research defended the interests of the US agrifood sector. If this represented an American 'head start', this would be brought to an abrupt halt in 1981 after the election of the neoliberal Ronald Reagan as US president. Now on the agenda was a massive reduction in federal budgets, starting with that of the Department of Energy. The climate programme was one of the first to pay the price.[81]

An independent institution, the AAAS was not itself directly affected. It pursued its climate-related activities for many years, though now with less wind in its sails. Thus in 1987 it organised a symposium in India on 'climate variations and food security'. Most important, the AAAS maintained a committee on the climate, chaired by Roger Revelle, which included atmospheric scientists such as Schneider, Kellogg and Orr Roberts, but also prominent economists like William Nordhaus and Thomas Schelling. A future Nobel Prize winner in economics, Schelling's importance would continue to grow – as we will see in chapter 2.[82]

INTERNATIONALISING ADAPTATION

Outside the United States, the climate question gave rise to multiple international initiatives for scientific cooperation. In 1967, the

80   'Climate Change and Agricultural Production in Non-Industrial Countries. A Report on the Results of a Workshop Held at The Institute for Energy Analysis, Oak Ridge, Tenn., 19–20 June 1980. Prepared for Department of Energy, Washington, D.C', Climate Project, AAAS, Washington, DC, 28 August 1980, pp. 17–18, AAAS Archives, box 3.

81   Howe, *Behind the Curve*, pp. 116–17.

82   See AAAS Archives, Series III, box 15.

International Council of Scientific Unions (ICSU), in collaboration with the World Meteorological Organization (WMO), developed a global programme for atmospheric research. In 1975, the WMO created an expert panel on climate change, and in 1979 organised the first global conference on the climate. This conference identified CO_2 as a crucial problem which demanded international research coordination; this, in turn, led to the creation of the World Climate Programme (WCP). Other scientific initiatives were also either launched or supported by the UNEP and UNESCO.[83] But as we have seen with the Aspen Institute and US Department of Energy programmes, what was under investigation in the late 1970s was not just climate science, but also the social, political and economic consequences of climate change – and the necessary political responses.

The International Institute for Applied Systems Analysis (IIASA) – a scientific institution that directly emanated from the Cold War – would contribute heavily to this vision. The fruit of the US authorities' desire, under Lyndon B. Johnson's presidency, to seek détente in East–West relations by way of scientific cooperation, the IIASA was established at Laxenburg Castle in Austria in 1972 and played host to scientists from both blocs.[84] Dedicated to cybernetic sciences, systems theory and modelling the world, the IIASA rapidly turned its attention to climate questions, and more particularly their social and economic dimensions – a specialism for which it is recognised to this day.

In 1978, the IIASA organised an international meeting on the impacts of climate change, especially with regard to energy systems.

83   Melinda L. Cain, 'Carbon Dioxide and the Climate: Monitoring and Search for Understanding', in David A. Kay and Harold K. Jacobson (eds), *Environmental Protection: The International Dimension*, Totowa, NJ: Allenheld Osmun, 1983, pp. 75–100.

84   The IIASA remains little studied despite this institution's crucial role in the development of the analysis of climate change and its consequences. One resource is Eglė Rindzevičiūtė, 'Toward a Joint Future Beyond the Iron Curtain: East–West Politics of Global Modelling', in Jenny Andersson and Eglė Rindzevičiūtė, *The Struggle for the Long-Term in Transnational Science and Politics: Forging the Future*, London: Routledge, 2015, pp. 115–43.

Cutting CO_2 emissions through controls on fossil fuels was considered a controversial proposal.[85] The economists called for cost/benefit calculations to be integrated into any policies that might be developed. One representative of the German coal producers' association remarked that given the uncertainty over climate change, the only urgent task was to build new power plants. That is, it was necessary to respond to energy needs rather than worry about the effects of global warming. Stephen Schneider called for a strategy of resilience, one able to respond to the ontological uncertainty of climate variability. That same year, the IIASA hosted a major international seminar designed to prepare the 1979 global climate conference. It spelled out the need to treat the climate as a resource and thus find a way to attribute it an economic value.[86] The seminar moreover underlined the need to live with climate uncertainty by developing new modelling methods and economic instruments.

Two years later, the IIASA continued its climate-related activities with a fresh conference on 'climatic constraints and human activities', bringing together several veterans of the Annapolis conference.[87] Basing his thesis on a study of droughts on the Great Plains of the United States, the geographer Richard Warrick showed that a society's capacity to adapt would allow it to mitigate the effects of climate variations.

Jesse Ausubel, one of the pioneers in the economics of climate change, discussed the difficulty of regulating access to the atmosphere, given both multiple competing types of usage and the absence of property rights over it. He called for the creation of a market in pollution permits. Acknowledging that it would be impossible to establish an

85 Jill Williams (ed.), *Carbon Dioxide, Climate, and Society: Proceedings of a IIASA Workshop, Cosponsored by WMO, UNEP, and SCOPE, 21–24 February 1978*, Oxford: Pergamon Press, 1978.

86 US National Research Council, Climate Research Board, *International Perspectives on the Study of Climate and Society*, Washington, DC: National Academy of Sciences, 1978.

87 Jesse Ausubel and Asit K. Biswas (eds), *Climatic Constraints and Human Activities: Proceedings of a Task Force on the Nature of Climate and Society Research, February 4–6, 1980*, Oxford: Pergamon Press, 1980.

international accord on reducing CO_2 emissions, he argued that it might be necessary to think about how to live with and adjust to 'an ever more irrationally exploited and degraded atmosphere'.[88]

The contribution by the German philosopher Klaus Meyer-Abich was little different. Out of the three possible political responses – prevention, compensation or adaptation – he considered only the last one realistic. The development of humans' capacity to predict the future climate would make it possible to reduce the costs of adaptation through a strategy of anticipation rather than a simple reaction to change. According to Meyer-Abich, this would principally take the form of several hundred million people migrating toward climatically better-off areas. But, he noted, 'I do not see any reason whatsoever why adaptation should not be the most rational political strategy'.[89]

BUILDING RESILIENCE

At the end of the 1970s, US climate research was at the cutting edge of research globally. But more than a simple matter of scientific curiosity, the climate question had become an existential problem for the US superpower. Research programmes sought not only to increase our understanding of how the atmosphere works, but also to evaluate the impact of climate change on human societies and their economies.

The mounting climate problem (whether this meant global cooling, or even more so, global warming) loomed over the future survival – and the legitimacy – of the 'free enterprise' system, the capitalism of which the US was both the guarantor and the main propagator globally. The impact of climate variations on human societies was generally read through neo-Malthusian lenses, which raised doubts over agriculture's capacity to feed the global population. There were

---

88   Jesse Ausubel, 'Economics in the Air: An Introduction to Economic Issues of the Atmosphere and Climate', in Ausubel and Biswas, *Climatic Constraints and Human Activities*, p. 55.

89   Klaus Meyer-Abich, 'Chalk on the White Wall?: On the Transformation of Climatological Facts into Political Facts', in Ausubel and Biswas, *Climatic Constraints and Human Activities*, p. 69.

insistent references to the possibility of a collapse – or at least a partial one. Some climatologists offered their services to US agrifood corporations to help these firms to emerge unscathed from this difficult period.

At the same time – paradoxically – many contributors, especially those from the social and human sciences, focused on human societies' capacities for long-term adaptation faced with all sorts of environmental transformations, including climatic ones. Adaptation was, in this presentation, a default option: what happened in society when changes occurred in the environment.

The main idea pervading these debates was the assumption that climate change was largely beyond the capacity of human societies to exert control. But not all societies would be equally vulnerable to these new shocks – and it would be possible to 'build resilience' to mitigate the effects. Strategies for deliberate adaptation, as championed by climatologist Stephen Schneider, dominated the debate.

Fundamentally, the possibility of reducing CO_2 emissions in order to limit climate change was perceived as politically too difficult, on an international scale – and much too costly for economies which were massively dependent on fossil fuels. In this context, proactive adaptation or building resilience appeared, from the US point of view, as the most reasonable path in response to the problem of climate change.

Chapter 2
The Gospel of Flexibility

Technologies are available for adaptation to climate on a spectrum of space, time and cost. Within minutes and for a few dollars one can buy an umbrella for local protection against a shower.

Jesse Ausubel, *Nature*, 1991.[1]

In November 1981, the popular science magazine *Technology Review* hosted a debate on the 'carbon dioxide problem'. The first of the two responses it published was not so surprising: it held that cutting CO_2 emissions was necessary in order to avoid dangerous global warming. Reducing the use of fossil fuels, starting with petrol, seemed like a reasonable solution. The transition to renewable energies was simultaneously desirable, possible and beneficial: the problem could be treated at source.[2]

The economist Lester Lave was a member of the AAAS committee for the study of carbon dioxide. And he was not ready to accept the idea that *reduction* was the best policy. Fossil fuels were necessary for the US economy, he pleaded, and it would be irresponsible to argue for doing without them when there was still such uncertainty over climate change.[3] Yet this did not mean that the climate question should be ignored. Rather, it had the great merit of forcing US society to introduce more flexibility into its modes of functioning:

---

1 Jesse H. Ausubel, 'Does Climate Still Matter?', *Nature*, vol. 350, 1991, p. 652.

2 Don Scroggin and Robert Harris, 'Reduction at the Source', *Technology Review*, vol. 84, 1981, pp. 22–8.

3 Lester B. Lave, 'The Carbon Dioxide Problem: A More Feasible Response', *Technology Review*, vol. 84, 1981, pp. 28–31.

Carbon-dioxide buildup can provide a rationale – but more probably it will be a catalyst – for enhancing society's ability to adapt to and exploit a changing environment. The issue is but one of many that will have an enormous impact on the world economy and social institutions in the twenty-first century, and it provides one more argument to make these institutions flexible, adaptable, and strong.[4]

Climate constraints had ceased to be an apocalyptic threat which would lead to a more interventionist state. For Lave, on the contrary, they justified a flexibilisation of society, in order to make it more adaptable to the relentless transformations of the contemporary world. It was no accident that such apologias for flexibility in response to climatic constraints surfaced at the beginning of the 1980s. For this marked the encounter between the (Malthusian) problematic of global limits on growth and the developing problematic of growing risks – and their management – in a world seemingly marked by uncertainty and constant change. Above all, these apologias were an expression of the counter-reform whose foundations the economic and political elites of the capitalist world were now laying. In promoting the new gospel of flexibility, the champions of adaptation to climate change contributed to the elaboration – and as we shall see in the following chapters, the practice – of the political ideology of neoliberalism.[5]

THE ECONOMISTS ENTER THE STAGE

The United States was a pioneer in environmental regulation. Progressive social movements forced the introduction of the main laws protecting natural resources and the environment in the 1960s

---

4   Ibid., p. 31.

5   I propose the idea of a gospel of flexibility with reference to Joan Martinez-Alier's classic work, which described environmental modernisation policies as a 'gospel of eco-efficiency' (Joan Martinez-Alier, *The Environmentalism of the Poor*, Cheltenham, UK: Edward Elgar, 2002) and to Howe (*Behind the Curve*, chapter 7) who writes about the 'market gospel', here describing the free-marketeer policy for governing the climate question, enacted from Rio onward.

and 1970s.[6] At that time, environmental policy consisted of an interventionist regulatory model founded on a substantive law. The legislator selected targets to be attained (for instance, tolerable norms of air pollution) and the state made these goals a practical reality by issuing the relevant regulations. And all this was crowned by powers of sanction. In short, environmental law drove the regulator-state to intervene ever more directly into companies' production decisions. The neoliberals feared that ecological demands would begin to weigh down the economic system and distort the functioning of the free play of prices – and all the more so, given that environmental questions were not the only domain affected. The combativity of the working class and struggles for civil rights were also undermining companies' freedom of operation.[7] The neoliberals feared that political decisions would prevail over market choice – that is, economic agents' decisions on what to pay.[8] For them, while government intervention was sometimes justified, it should take place in forms compatible with the market. Future Nobel Prize laureate in economics Thomas Schelling wrote in 1979 that 'the issue is not governmental intervention on behalf of the environment so much as it is the mode of intervention, the philosophy of costs and benefits, and the locus of decision'.[9]

In this context, some economists rapidly came to see the climate question as a threat. But what they most feared was not the threat of climate catastrophe, but rather the threat of extended state regulation. For Schelling:

Environmental protection is often treated, officially as well as popularly, as an absolute – not as an economic choice, not as a correction

6   Dahan and Aykut, *Gouverner le climat?*, chapter 4.

7   Panitch and Gindin, *The Making of Global Capitalism*, chapter 4.

8   Benjamin J. Richardson and Stepan Wood (eds), 'Environmental Law for Sustainability', in *Environmental Law for Sustainability*, Oxford: Hart, 2006, pp. 1–18; Romain Felli, 'Environment, Not Planning: The Neoliberal Depoliticisation of Environmental Policy by Means of Emissions Trading', *Environmental Politics*, vol. 24, no. 5, 2015, pp. 641–60.

9   Thomas Schelling, Committee on Long-Range Energy Policy, *Thinking Through the Energy Problem*, New York: Committee for Economic Development Design, 1979, p. 56.

applied to the price system, not even as part of the cost of our energy, but as a matter of regulatory standards and prohibitions to be judged and administered without compromise, sometimes as a kind of militant opposition to economic improvement and growth.[10]

The first US economists who studied climate change, like Thomas Schelling, William Nordhaus, Mancur Olson, Jesse Ausubel, Gary Yohe and Lester Lave, were highly sensitive to the threat of regulation. But they also had a developed sense for the ways to deflect such regulation, in order to make sure that it did not affect capital accumulation. In a context in which revanchist neoliberal governments had come to power in both Britain (Margaret Thatcher, in 1979) and the United States (Ronald Reagan, in 1981) their arguments would have decisive influence.

The economists' response to the regulatory threat posed by the climate question operated at two levels. First, they sought to relativise the climate problem, making it comparable to other economic questions: through cost/benefit analyses, the climate question could be treated not as a moral question but rather as a choice of how to allocate resources. Second, once climate policies seemed to have become an inevitability, the economists worked to ensure that these policies would conform, as far as possible, to the operation of the market. Together, these two responses implied the need to emphasise the theme of adaptation.

COSTS OF MITIGATION, BENEFITS OF ADAPTATION

The first strategy demanded a comparison between the costs of reducing greenhouse gas emissions and the costs of the predicted effects of global warming. If this comparison was itself complicated, Schelling proposed that it should be combined with a comparison of the costs and benefits of different possible responses to the problem. The comparison would have to take into account not only techniques for

---

10 Ibid., pp. 54–5.

decarbonising the atmosphere (for instance, planting trees, whose growth would absorb CO_2) but also techniques for the deliberate modification of the climate and techniques for adaptation.[11] This line of thinking was part of the rapid rise of cost/benefit analysis within US public policy, as notably applied to the environment during the 1970s.[12]

If warming reduced water availability in a given region, the economist ought to compare the costs of reducing greenhouse gas emissions with the costs of producing extra water, for instance by building a desalination facility. Similarly, if it was less expensive to absorb CO_2 by planting trees than by reducing petrol usage, then this first solution should be preferred. Conversely, 'deciding how much CO_2 suppression makes sense requires an assessment of adaptation in the aggregate.'[13] The distribution of costs and benefits, thus calculated, became crucial. The reduction of fossil fuel use would hurt producer countries (the USA, USSR and China, in the case of coal), while the climate benefits would apply to the entire planet. Conversely, the benefits of adaptation measures (building dams, cultivating plants resistant to variations of temperature, etc.) would flow back to those

---

11   Thomas Schelling, 'Anticipating Climate Change', *Environment*, vol. 26, no. 8, 1984, pp. 6–9 and 28–35. This article reproduced Schelling's contribution to the US National Research Council report, *Changing Climate: Report of the Carbon Dioxide Assessment Committee*, Washington, DC: National Academy Press, 1983. The historians Naomi Oreskes, Erik M. Conway and Matthew Shindell ('From Chicken Little to Dr. Pangloss: William Nierenberg, Global Warming, and the Social Deconstruction of Scientific Knowledge', *Historical Studies in the Natural Sciences*, vol. 38, no. 1, 2008, pp. 109–52) portray this report and its author, Bill Nierenberg, as a major source of the 'deconstruction' of scientific knowledge on global warming and thus of contemporary climate scepticism. They point out that Nierenberg accepted the arguments coming from economists (Thomas Schelling, William Nordhaus, Jesse Ausubel, Gary Yohe) who tended to play down the problem of climate warming and celebrate society's capacity to adapt, to the detriment of the need to reduce greenhouse gas emissions. While I do not disagree with their more general analysis, it seems to me that these historians strongly underestimate how far the theme of adaptation had spread in climate debates well before this 1983 report – and far beyond economists' own circles.

12   Soraya Boudia, 'Gouverner par les instruments économiques: La Trajectoire de l'analyse coûts-bénéfices dans l'action publique', in Pestre (ed.), *Le Gouvernement des technosciences*, pp. 231–59.

13   Schelling, 'Anticipating Climate Change', p. 31.

who pursued these measures. This was to their great economic merit, as it got rid of the 'free-rider' effect (so dear to Mancur Olson), a drawback of voluntary reductions in CO_2 emissions.

Schelling also pointed out that global warming did not have only negative effects. In Canada and the United States, farmers would gain the ability to use land currently uncultivated because of longer cold winters; winter energy bills would also fall, etc. And it should be added that this analysis was widely shared. In 1978 the climatologist William Kellogg could write that, thanks to global warming, the earth would have a climate generally more favourable to feeding a growing population.[14] Subsequently, the little research funding that existed for climate studies during the Reagan years was funnelled toward research detailing the positive consequences of warming.[15]

But the economic reading of the climate question was not the sole preserve of economists – especially as concerned comparisons of the relative costs of different political options. A climatologist like Schneider was not alien to this way of thinking, and nor was an ethical philosopher like Meyer-Abich.[16] The conclusion to which this led in the early 1980s was that, given the uncertainties which still existed, but in particular the unequal distribution of the costs and benefits of the various possible measures, it 'ma[de] sense, therefore, to anticipate changing climates'.[17]

The neoliberal economists' first response to the climate threat thus consisted of playing down the need to reduce greenhouse gas emissions. Even where such reductions would be necessary, they pleaded for them to be enacted through 'economic instruments' which would afford capital the greatest flexibility: voluntary

---

14   William Kellogg, 'Global Influence of Mankind on the Climate', in John Gribbin (ed.), *Climatic Change*, Cambridge: Cambridge University Press, 1978, p. 223.

15   Howe, *Behind the Curve*, p. 123.

16   Schneider and Chen, 'Carbon Dioxide Warming and Coastline Flooding'; Meyer-Abich, 'Socioeconomic Impacts of CO_2-Induced Climatic Changes'. More generally, this worldview would be replaced by an insistence on evaluating the environment in economic terms: see Dominique Pestre, 'Néolibéralisme et gouvernement', in Pestre (ed.), *Le Gouvernement des technosciences*, pp. 261–84.

17   Schelling, 'Anticipating Climate Change', p. 35.

measures, private certification, environmental taxes, markets in pollution permits and so on. The 'polluter pays' principle developed within the OECD was at the heart of this doctrine – you can pollute, so long as you are able to pay. Thus depoliticised, environmental legislation would allow businesses greater freedom of manoeuvre.

THE POLITICS OF CLIMATE UNCERTAINTY

What most interests us here is the history of the economists' second response. It involved a reframing of the debate, taking it further away from the question of CO_2 emissions or worse – in their eyes – fossil fuels as a whole. For Schelling, the debate 'should be built around climate change, not around CO_2.'[18] Posed in these terms, the task was not to combat a given practice (the use of fossil fuels) in order to avoid climate change, but rather to learn to live with this change. Yes, climate change was very real. But, the argument went, it was just one of the many changes affecting society: technological revolutions, changing moral codes, the offshoring of jobs, urbanisation, etc. These constant changes created a situation in which, according to Schelling, there reigned but 'one certainty: uncertainty'.[19]

In this regard, it is worth distinguishing between two forms of uncertainty which the climate question brings into play. Even if experts consider it impossible to provide any detailed prediction of the future climate, they share the idea that with a lot of work – and resources – it would be possible to improve the accuracy of climate modelling. But the first uncertainty has to do with climate change itself: Is it real? What is its scale? What are its origins? How will it evolve? This is, then, an epistemological uncertainty, linked to a knowledge gap. This uncertainty is an annoyance, but it is also temporary and reducible.

Yet the works of the 1970s and early 1980s emphasised a second form of uncertainty, *created by* climate change. Researchers

18   Ibid., p. 8.
19   Schelling, *Thinking Through the Energy Problem*, p. 61.

demonstrated the rising variability of the climate, or of its fluctuations, in a given region. In other words, one of the major consequences of climate change was that it led to a climate that varied more than it had in the past. For instance, the experts who convened at Bellagio in 1975 stated that

1. Climatic variability – region by region and from year to year in particular regions – is and will continue to be great, resulting in substantial variability in crop yields in the face of increasing global food needs and short supplies;

2. There is some cause to believe – although it is far from certain – that climatic variability in the remaining years of this century will be even greater than during the 1940–1970 period.[20]

This second uncertainty was ontological in nature: the world (or the climate, at least) would become more changeable and thus structurally less predictable from one year to the next, in any given region.[21] This second uncertainty was especially attention-catching because, unlike the first, it seemed more difficult, or even impossible, to reduce. For many observers, climate change, variability, and thus uncertainty, were a new and decisive reality which had to be integrated into the way in which nature and society were managed. Climate politics became a politics of uncertainty.

---

20 Bellagio Conference, 'Climate Change, Food Production, and Interstate Conflict', pp. 33–4.

21 This thesis was disputed by other researchers at the time, for whom the only thing that could be measured was the increased perception of variability by an ever-better-informed public (an epistemological change) and not a real increase in climate variability. See, for example, Jim Norwine, *Climate and Human Ecology*, Houston: D. Armstrong, 1978, p. 233.

ADAPTATION AS ADJUSTMENT TO NATURAL RISKS

One first way of anticipating these fluctuations – and creating the required flexibility in response – was advanced by geographers specialising in natural risks. If their contributions cannot be equated with the economists' neoliberal proposals, they nonetheless remain important. Their approach called for behavioural transformation in order to achieve the resilience necessary to cope with the incessant transformations of society and the environment.

The experts brought together by the Canadian climatologist Kenneth Hare in Toronto in 1978, under the auspices of the ICSU's SCOPE programme, noted that 'resilience refers to the ability of a society to "bounce-back" when adversely affected by climatic impacts'.[22] While catastrophism tends to cast climate change as a global threat to human societies, their analysis counterposed different societies' capabilities to react to these changes. Again, it was necessary to acknowledge and strengthen this capacity:

> considerable opportunity may well exist to reduce climatic impacts and to increase resilience; in a minor way through attempts to modify or control climate; in a major way through a clearer understanding of the pattern and character of development activities themselves.[23]

The effects of climate change were to be understood not by studying the climate, but by studying societies – that is, studying their capacities to anticipate risks and react to them. These impacts, and the ability to react to them, are distributed unequally within societies: the poorest are often more exposed to the effects of climate change and less able to cope with them. And social reactions that could increase resilience also risk undermining it if the chosen solutions worsen the problem.

---

22 ICSU/SCOPE, *ICSU/SCOPE Workshop*, p. 15.
23 Ibid., p. 16.

The report from the Toronto seminar was drawn up under Hare's direction by a group including a young geographer called Ian Burton. He had already been working for some years on natural risks and the social consequences of climate risks. Along with his PhD supervisor Gilbert F. White – a US geographer and pioneer in the social analysis of natural disasters[24] – and Robert Kates, Burton in 1978 published a short book which was destined to become a point of reference: *The Environment as Hazard*.[25] This book provided a synthesis of the vast research programme pursued over previous years at the Universities of Chicago and Toronto as well as Clark University (Massachusetts). This tradition of thinking paved the way for the recommendations of the SCOPE seminar.

Burton and his colleagues analysed the way in which societies and individuals prepared for and responded to natural hazards – for example, the risk of an earthquake or a period of drought. Their works showed that the impact of a natural hazard depends less on Nature than on the way that societies anticipate and respond to such hazards. For instance, a local authority that stocks grain reserves can prevent drought turning into famine; buildings constructed according to anti-seismic norms will withstand an earthquake where a conventional one would collapse; an effective health system and civil protection will stop epidemics spreading after a flood; and so on.

These geographers distinguished between adaptation and adjustment. The adaptation of societies to their environment was termed a biological, or cultural, process – it took place over the longue durée and had to do with unconscious processes. A Darwinian approach underpinned this notion of adaptation, in this era mobilised by cultural anthropology to take account of the supposed symbiosis

24 On White and the Chicago school of analysis of natural disasters, see Magali Reghezza's entries 'École de géographie de Chicago', 'White, G. F.' and 'Adaptation' in Yvette Veyret (ed.), *Dictionnaire de l'environnement*, Paris: Armand Colin, 2007. White long played a role as a front-rank political advisor, including as part of the national water policy commission set up by President Truman in 1950. This commission was one of the sites where the cost/benefit analysis of environmental issues was developed: see Boudia, art. cit. p. 235.

25 Ian Burton, Robert W. Kates and Gilbert F. White, *The Environment as Hazard*, Oxford: Oxford University Press, 1978.

between 'traditional' societies and their environment.[26] But for Burton adaptation stood beyond the scope of risk studies, for it could not be deliberately modified. On the contrary, *adjustments*, whether conscious or unconscious, were inscribed in the short term and could indeed be the object of politics.

Influenced by risk psychology and the nascent behavioural economics (notably that of Daniel Kahneman), Burton and his colleagues analysed the choices that individuals make when faced with environmental hazards. In their account, these are decisions based on cost/benefit analysis. For example, what would it cost a fisherman to find shelter for his boat in response to a typhoon warning (the effort of taking it out of the water, the time lost not fishing, etc.) relative to the probability that it is a false alarm? In this perspective, one's degree of risk exposure is a choice.[27] Unfortunately, individuals do not have complete information available to them when they make their choices. Their rationality is thus 'limited'. It is the public authorities' job to increase the quantity and quality of information to help guide citizens' choices on exposure to risk. For instance, the building of weather stations contributes to this reduction of uncertainty, albeit within a brief temporal horizon.

These recommendations, which seek to alter individual behaviour through public incentives, systematised the thesis which White had forged through his study of US flood prevention policies in the 1930s.

White was quick to express his scepticism regarding attempts to reduce risk solely by means of hazard control. In damming the rivers, the

---

26   For instance, John W. Bennett, 'Anticipation, Adaptation, and the Concept of Culture in Anthropology', *Science*, vol. 192, no. 4242, 1976, pp. 847–53; *The Ecological Transition: Cultural Anthropology and Human Adaptation*, New York: Pergamon Press, 1976. Michael J. Watts provides a very detailed genealogy of this functionalist way of thinking (among authors such as Roy Rappaport), which he situates at the conceptual origins of the present-day policies of adaptation. See his important article, 'Now and Then: The Origins of Political Ecology and the Rebirth of Adaptation as a Mode of Thought', in Tom Perreault, Gavin Bridge and James McCarthy (eds), *The Routledge Handbook of Political Ecology*, London: Routledge, 2015, pp. 19–50.

27   Burton, Kates and White, *Environment as Hazard*, p. 11.

authorities had given a false impression of control and security: populations made their homes behind dams which, if they gave way, would create a flood causing much greater damage than if the water had been allowed to flow naturally. White favoured a national (public) system of flood insurance, which would allow the victims to be compensated and to rebuild their properties. The counterpart to this insurance would come through territorial planning measures which would forbid or limit construction in at-risk areas. Those 'incautious' types who persisted in building in flood zones despite this signal from the authorities would thus see their premiums jacked up or their compensation slashed.

Over the decades that followed the Toronto seminar, Burton established himself as one of the world's leading specialists on adapting to climate change. He abandoned his reticence regarding the Darwinian undercurrents of the notion of adaptation and became an influential consultant for the World Bank and one of the leading authors of the IPCC reports dedicated to this question.[28] His approach sought to increase societies' capacity for resilience by reducing the 'deficit in adaptation' to climate change. The notion of adapting to climate change which was forged by international institutions and reproduced in the bulk of the scientific literature was strongly influenced by this analysis. At first glance, it offered a message of hope: for in modifying their institutions and their behaviours, societies could overcome the shock of climate change.

However, this functionalist vision of adaptation as adjustment tends to ignore the deeper causes of vulnerability to the climate. In the 1990s and 2000s it nonetheless oriented both scientific research and international expertise (especially in the IPCC) on adaptation.[29]

---

28   Some of his contributions appear in a collection he co-edited: E. Lisa F. Schipper and Ian Burton (eds), *The Earthscan Reader on Adaptation to Climate Change*, London: Earthscan, 2008. See also Ajay Mathur, Ian Burton and Maarten Van Aalst (eds), *An Adaptation Mosaic: A Sample of the Emerging World Bank Work in Climate Change Adaptation*, , Washington, DC: World Bank, 2004.

29   Thomas J. Bassett and Charles Fogelman, 'Déjà vu or Something New?': The Adaptation Concept in the Climate Change Literature', *Geoforum*, vol. 48, 2013, pp. 42–53; Jesse Ribot, 'Vulnerability Before Adaptation: Towards Transformative Climate Action', *Global Environmental Change*, no. 21, 2011, pp. 1160–2.

While it seeks to improve societies' capacities for adaptation, it invisibilises the social relations, especially economic relations, which structurally render certain parts of the population vulnerable.[30] Some contemporary researchers counterpose this 'resilient' vision of adaptation with a 'transformative' one that does not settle for trying to increase adaptation capacity but rather aims to transform the social relations that produce vulnerability.[31]

THE UGLY DUCKLING: THE GARCÍA EPISODE

Despite its intentions, the Aspen seminar in 1973 would draw increasing attention to societies' vulnerability to environmental changes. The famine in the Sahel was at the heart of its discussions. The programme directed by Walter Orr Roberts at the Aspen Institute linked up with the International Federation of Institutes for Advanced Study (IFIAS) to finance a far-reaching study of the causes of famine: *Drought and Man: The 1972 Case History*. Entrusted in 1976 to the Argentinian climatologist Rolando V. García, who brought together a large interdisciplinary team of researchers, its conclusions were published in three volumes starting in 1981.[32] A physicist specialising in the atmosphere, García was also a philosopher of science. At the moment he was appointed to head up the IFIAS study, he was a close collaborator of Jean Piaget at the University of Geneva.

Contrary to the views of those who had commissioned his report, García radically broke with the methodology of climate 'impact studies'. It was expected that he would present the climatic

---

30 This critique is advanced by the radical analysis of disasters, which made up part of the origins of the analytical current today called political ecology. See the classic book edited by Kenneth Hewitt, *Interpretations of Calamity from the Viewpoint of Human Ecology*, Boston: Allen & Unwin, 1983 (especially the contributions by Hewitt and Watts); and Ben Wisner, Piers Blaikie, Terry Cannon and Ian Davis, *At Risk: Natural Hazards, People's Vulnerability and Disasters*, 2nd edn, London: Routledge, 1994.

31 Pelling, *Adaptation to Climate Change*.

32 Rolando V. García, *Drought and Man: The 1972 Case History*, Oxford: Pergamon Press, 1981–1986; vol. 1: *Nature Pleads Not Guilty*, 1981; vol. 2: *The Constant Catastrophe: Malnutrition, Famines, and Drought*, 1982; vol. 3, *The Roots of Catastrophe*, 1986.

and physical conditions of the problem before turning to the trig-
gering event (the drought) producing disastrous effects (the
famine), and then to the societal responses (adaptation). But
García's study turned this determinist vision on its head: it began
by examining the social conditions in which the famine had taken
place. The question of climatic variability was limited to a contribution by the US climatologist Joseph Smagorinsky, which was rele-
gated to an annex. The brilliant title of this provocation? *Nature
Pleads Not Guilty*!

Far from bolstering the neo-Malthusian narrative that located the
origins of the Sahel famine at the crossroads between drought and
population increase, García instead situated its roots in the political
economy of the region and the resulting socioeconomic vulnerabilities. The book's introduction attacked that research which cast society
as a simple receptacle for the 'impacts' of the climate system and
developed 'adaptive responses' in return.[33] Within a few months of
beginning his study, García understood that the drought had played a
secondary role in the famine being analysed: at most, it had triggered
a calamity which was in fact social in origin. He demolished the positions of the likes of Lester Brown and Roger Revelle on overpopulation and the supposed fall in agricultural production owing to climate
variations. He even pointed to *depopulation* as a problem for the
Sahel. The rise in grain prices owing to speculation on international
markets, the development of an agriculture aimed at export rather
than feeding the population, bad economic policy choices, the lack of
social security and, more generally, the collapse of the socio-economic
system in the Sahel had combined in producing extreme vulnerability
– and exposing certain populations to it. The famine was not a natural
disaster but a social one.

García concluded 'that structures – i.e. social organization; political systems; control of credit and capital; large-scale production
arrangements, such as the multinational agro-business; international
inter-governmental organizations and their bureaucracies' determined

---

33   García, *Drought and Man*, vol. 1, *Nature Pleads Not Guilty*, p. xiii.

access to food.'[34] To attribute malnutrition and famines to climate variations alone would, conversely, mean denying this disastrous economic and political situation. The study's conclusion went on the offensive against generalising predictions about the impacts of climate change – indeed, it ridiculed them. Only a concrete analysis of the vulnerabilities of different socio-economic systems could make any sense. In short, 'we should not fool ourselves into believing that we are really assessing the possible effects of an increase in CO_2.'[35] Worse, attempts to 'adapt' to climate change which ignore the conditions that produce social vulnerability could only lead back to the initial state of poverty – the very source of the problem – and not to the transformation of these conditions.

The neo-Malthusian climatologists who had commissioned this study certainly did not appreciate its results. Walter Orr Roberts's foreword thundered that García's arguments were those of the author alone and in no way reflected the position of the IFIAS or the Aspen Institute. He called for a 'critical' evaluation of its 'controversial' theses, and for the posing of other points of view in opposition to García's own. In turn, in his acknowledgements García diplomatically thanked his funders, even though his study 'might not be liked by some of the sponsoring institutions'.[36]

Grounded in an analysis of socio-economic vulnerabilities and political economy, García's work exploded the determinist narration of the climate question. Not surprisingly then, García's position was marginalised when he was invited to the IIASA seminar in 1978 to present the results of his research. The results of his presentation were provided as an appendix but ignored in the seminar report.[37] The same went for the three volumes of *Drought and Man*, which rather rapidly sank into a polite neglect.

---

34   Ibid., p. 223.
35   Ibid., p. 221.
36   Ibid., pp. v–vii.
37   See National Research Council, *International Perspectives on the Study of Climate and Society*, 1978, especially Rolando García, 'Annex C: Climate Impacts and Socioeconomic Conditions', pp. 43–7.

The social analysis of famines and natural disasters enjoyed an impressive rise at the beginning of the 1980s. The Indian philosopher and economist Amartya Sen made his name with his study of the relations between political power, economic power and access to food.[38] Marked for life by his childhood experience of the great Bengal famine in 1943/44 – during which 3 million to 5 million people died! – Sen demonstrates that malnutrition and famines are rarely caused by an absolute shortage of food. Taking the example of the Sahel famine, he pointed to the unequal distribution of social rights as the cause of unequal access to food. Famine is the result of political choice, not the wickedness of Nature. The tragedy in the Sahel also featured at the heart of another major contribution to the relations between society and environment, this time by the Marxist geographer Michael Watts.[39] Challenging the neo-Malthusian reading of the famine in the Sahel, he produced a longue durée history and anthropology of the social formation in northern Nigeria and of the production of social vulnerabilities to climate fluctuations. In particular, this history exposed the role of colonialism and the penetration of capitalist social relations.

These different analyses were theoretically and politically diverse. But they all pointed to the need to transform political and economic structures (for instance by allotting effective, non-market-dependent rights to individuals) in order to reduce social vulnerability, rather than merely dump the blame on climate variations or on the fertility of the soil.

According to Naomi Oreskes and Erik Conway's account, from the early 1980s the mercenaries of the oil industry systematically succeeded in sowing doubts over research into human-caused climate change – and did so through efforts aimed at a 'deconstruction of knowledge.'[40] But if this deconstruction undeniably took place, that is

---

38   Amartya Sen, *Poverty and Famines: An Essay on Entitlement and Deprivation*, Oxford: Oxford University Press, 1981.

39   Michael Watts, *Silent Violence: Food, Famine and Peasantry in Northern Nigeria*, Athens: University of Georgia Press, 2013 [1983]; also Watts, 'On the Poverty of Theory'.

40   Oreskes, Conway and Shindell, 'From Chicken Little to Dr. Pangloss', 2008.

hardly the end of the matter. For the climate question was also marked by a 'marginalisation of knowledge'. The contributions of Rolando García, Michael Watts and Amartya Sen to the study of social vulnerabilities leading to famine were never integrated into climate impact studies. Rather, they were marginalised in favour of a generalising and determinist approach to climate change, dominated by global Earth system science.

THE GLOBAL SOUTH GATECRASHES TORONTO

But a challenge to climate determinism had now been posed, one that would have at least some bearing on the first international political conference on climate change, 'The Changing Atmosphere', held in Toronto in June 1988. This would no longer be simply a scholarly affair. Within the US government the climate question was still being discussed in terms of adaptation policies, but the Toronto conference now insisted on the need to cut greenhouse gas emissions. However, the Global South countries represented at the conference feared that this new emphasis on protecting the atmosphere as a 'common good' might divert attention from the more immediate problems of development – and serve as a pretext for neo-colonial encroachments upon their sovereignty.

Norway's social democratic prime minister, Gro Harlem Brundtland, opened the conference with a stirring call for the development of renewable energy, the pooling of clean technologies and strengthened international action to fight against the warming of the planet.[41] Her message was clear: economic growth was necessary to find solutions to environmental – and particularly climate – problems, on the condition that it was well managed. Garlanded by the international echo of the report by the World Commission on Environment and Development, which Brundtland herself chaired,

---

41 *Conference Proceedings: The Changing Atmosphere: Implications for Global Security. Toronto, Canada, 27–30 June 1988*, WMO no. 710, Geneva: Secretariat of the World Meteorological Organization, 1989, pp. 16–30.

her speech drew close attention. But as Kenneth Hare later noted, Brundtland's *Our Common Future* – later the 'bible' of sustainable development – did not really address global warming.[42]

The tone Brundtland adopted was echoed by all the main speakers: a mix of wonderment at the enormity of the global change driven by global warming combined with unfailing optimism in the possibility of reaching an international agreement to cut greenhouse gas emissions. Even while admitting the difficulties that some coastal and island states would encounter, the UNEP representative seemed confident in the possibility of reaching an international agreement that would allow not only preparedness in the face of climate change but even 'drawing as much benefit from it as possible'.[43] However, the 'action plan' adopted by the conference concerned only measures to cut greenhouse gas emissions. As for international funding – something to which serious reference was now being made – the plan included only a relative reduction in Global South countries' emissions, not efforts to finance a reduction in their vulnerability to global warming.

The report's one-sided focus on cutting emissions did not, however, reflect the work that had been done at the conference. While all participants shared the desire to reduce emissions, many did not want to deal with the climate question in isolation from development. For instance, the working group on food security demanded the immediate enactment of resilience measures (building up stocks, improving infrastructure, etc.) in order to reduce populations' vulnerability to climatic variability.[44]

The participants from Global South countries insisted on challenging the narrative that presented warming uniformly as a global threat. The Indonesian environment minister, Emil Salim, noted that countries had different capacities to respond to atmospheric changes, and that these differences were largely determined by their levels of

---

42   Hare, in 'The Changing Atmosphere', pp. 59–69.
43   William Mansfield III, in 'The Changing Atmosphere', p. 275.
44   Ibid., pp. 321–23.

wealth and development.[45] The climate was not responsible for the dangerous urbanisation in the countries of the Global South. Argentinian delegate Jorge Hardoy commented that the lack of health-care infrastructure, inadequate water and electricity networks and precarious construction taking place on hazardous terrain were each the result of the systematic exclusion of the majority of the population from access to wealth and political power. He asked whether, rather than trying to control the global atmosphere, it might not make more sense to deal with the direct – economic and political – causes of the vulnerabilities among indigenous, poor and excluded populations.

To discuss how to prevent the worst impacts of natural disasters on urban societies is to respect people, to become aware of their needs and to understand the processes that are building and rebuilding settlements day after day. In nations hit by economic crisis and recession, essential urban investments cease, real salaries decline, and the shortage of jobs, housing, social services and public utilities hits ever-increasing numbers.[46]

Above all, the way in which Global South delegates began to talk about adaptation differed in tone from the gospel of flexibility. If they recognised rising climate variability, driven by climate change, they did not conclude from this that it was necessary to flexibilise society and its institutions. On the contrary – in the same vein as the Polanyian counter-movement – Indian agronomy experts, for example, called for stabilisation measures. There needed to be political stability in order to cope with climate variability. The food policies of each state had to be known, fixing long-term targets and protecting producers over time by guaranteeing them sufficient income independent of the success of harvests from year to year. 'Whereas the effects of climate variability are difficult to anticipate and control', they write, 'the

45   Emil Salim, in 'The Changing Atmosphere', p. 161.
46   Jorge Hardoy, in 'The Changing Atmosphere', p. 237.

resulting instabilities may be minimized through appropriate food policy'.[47] To that end, grain reserves would have to be established, but in addition technical, scientific and administrative infrastructure that would support producers: road networks, distribution networks, technical education, etc. In short, long-term political planning was necessary in the face of the uncertainty driven by climate change.

THE ECOLOGICAL GOSPEL OF FLEXIBILITY

Yet the solution proposed by the champions of flexibility – especially the American ones – was exactly the opposite. Rather than extending social security, which would allow for greater stability and better predictability in troubled times, the ecological gospel of flexibility as preached in Lave's interventions instead proposed that permanent change was something to be embraced. This meant accepting instability and unpredictability in order to become more flexible, and thus better able to adapt.

The notion of flexibility, which emerged in the 1970s as a means of responding to environmental and especially climatic challenges, took on many different senses. At a relatively technical level, it simply signalled that physical infrastructure ought to be able to absorb extraordinary disturbances: for instance, dams should be designed in such a way that they would not only withstand a river's usual flow but continue to do so even in the case of a 100-year flood. Similarly, when Steve Schneider called for flexible agrifood systems in order to cope with the climate threat, in his mind this was a matter of building up reserves to absorb shocks (famines). And Schneider moreover considered this to be government's job, because the markets, left to themselves, would be incapable of bringing about security in food reserves.[48]

Schneider's remarks also remind us that 'flexibility' has no absolute political value – no more than does 'stability'. In the 1950s and

---

47   S. K. Sinha, N. H. Rao and M. S. Swaminathan, in 'The Changing Atmosphere', p. 181.

48   Schneider, *Genesis Strategy*, pp. 252–5.

1960s, conservative theorists championed stability while critical theorists, who denounced the one-dimensionality of society and its reproduction, instead valorised change, transformation and experimentation. But in the context of the neoliberal counter-reformation of the 1970s, attacks on the welfare state, bureaucracy and trade-union power inverted this critique. They now turned to the rhetoric of competition, flexibility and change – as against the 'blockages' in society.[49]

The ecological gospel of flexibility would henceforth proceed in this same direction. Those who preached this gospel identified the stability of social institutions – and their determination to control the human environment – as the main source of ecological problems and as an obstacle to successful adaptation. The determination to achieve a generalised flexibilisation of society – and its model, the market – now appeared as a solution to the climate question.

Thus, despite his abstruse prose – and a worldview bordering on New Age philosophy – the anthropologist and psychologist Gregory Bateson became an inevitable reference point for debates on adaptation.[50] Using the language of cybernetics and systems theory, over the 1960s and 1970s Bateson elaborated a highly functionalist approach to the relations between society and nature.[51] He denounced the triptych of overpopulation, technological developments and Western thought. In this synthesis, he showed no great originality.[52]

What Bateson then proposed as a means of restoring the 'health' of the environment was to increase the flexibility of human civilisation. If it was sufficiently flexible – and not governed by centralised political power – society could adapt as its environment changed. In this way, the two would make up a single complex system, in which

---

49   Michel Crozier, *La Société bloquée*, Paris: Seuil, 1970.

50   Watts, 'On the Poverty of Theory', and Pelling, *Adaptation to Climate Change*, p. 27.

51   Gregory Bateson, 'Effects of Conscious Purpose on Human Adaptation' (1968), in *Steps to an Ecology of Mind: A Revolutionary Approach to Man's Understanding of Himself*, New York: Ballantine, 1972, p. 446.

52   Bateson, 'The Roots of Ecological Crisis' (1970), in *Steps to an Ecology of Mind*, pp. 488–93.

human civilisation would take only minimal resources from the natural world, and for anything else would rely on solar power.[53] To realise this programme in just a few generations, 'very great *flexibility* will be needed',[54] with fexibility being defined as 'un-committed potentiality for change'.[55] Yet as Bateson saw things, society seemed incapable of such flexibility. It instead prepared to accept the problem (for example, by building more housing for a growing population) and reject classic Malthusian brakes (combatting epidemics and famines, rather than letting them do their work). It was thus exhausting its margins of flexibility. The ecologist's role was thus to combat the exorbitant demands of a population in search of security – and this would be for its own good, even if it meant showing a 'tyrannical' side.[56]

The 'planner' had to create a careful balance between the parts of society which had to be made more flexible, and those others whose flexibility should instead be restricted. If, in an early period, a sliver of security was necessary so that one could learn from one's mistakes, that element of security would later constitute a 'corset', restricting development. Incidentally, this was also the position of the OECD expert David Norse, who examined the impacts of climate change on poor peasants in the Third World. He claimed that a peasant who had too little protection from economic and climatic hazards would be incapable of taking the risks needed to enter and compete in the market.[57]

As for Bateson, he thought that the United States was a society whose flexibility had been unduly restricted. He instead suggested that the ecologist planner who wanted to build this 'flexibility

---

53   Bateson, 'Ecology and Flexibility in Urban Civilization' (1970), in *Steps to an Ecology of Mind*, pp. 494–5.

54   Ibid., p. 496.

55   Ibid., p. 497.

56   Ibid.

57   David Norse, 'Natural Resources, Development Strategies and the World Food Problem', in Biswas and Biswas (eds), *Food, Climate, and Man*, pp. 12–51. In this period, Norse was part of the OECD's 'Interfutures' project, a study which – critiquing the Club of Rome report – sought ways to maintain economic growth by 'managing unpredictability'. See Matthias Schmelzer, 'The Crisis Before the Crisis: The "Problems of Modern Society" and the OECD, 1968–74', *European Review of History*, vol. 19, no. 6, 2012, p. 1011.

economy' must continually press onward – for habits would immediately take root and make the system rigid again. Bateson cited a Zen
master who had supposedly told him that 'to become accustomed to
anything is a terrible thing.'[58]

Flexibility had to be 'exercised, or the encroaching variables must
be directly controlled.' According to Bateson, our societies had chosen
the second strategy, with over-the-top legislation. But the more desirable strategy was the former: 'It might be more effective to encourage
people to know their freedoms and flexibilities and to use them more
often.'[59]

Bateson addressed himself to planners (and especially urbanists)
in order to integrate this ecological thinking into their plans and to
influence subsequent development. The aim was that this thinking
should be fully integrated into society; and by doing so ecology would
become a science of social flexibility. As against institutions, which
produced an only illusory stability and security, Bateson preached
permanent change, and especially the integration of this flexibility
into behaviours, plans and projects, till they became part of our own
selves. 'Herein lies the charm and the terror of ecology – that the ideas
of this science are irreversibly becoming a part of our own ecosocial
system.'[60] Indeed, this was quite a 'revolutionary' approach!

RESILIENT FLEXIBILITY

Ecology, as a science of social flexibility, was destined to achieve striking success in the form of the theory of resilience. This theory was
especially widely used in the late 2000s in orienting international
policies for adaptation to climate change.[61] It would be wrong to

58   Gregory Bateson, 'Ecology and Flexibility in Urban Civilization', p. 503.
59   Ibid.
60   Ibid., p. 504.
61   Neil Adger, Katrina Brown and James Water, 'Resilience', in John Dryzeck,
Richard Noorgard and David Schlosberg (eds), *Oxford Handbook of Climate Change and
Society*, Oxford: Oxford University Press, 2011, pp. 696–709. For a literature review, see
Felli, 'Adaptation et résilience'.

present any narrow equivalence between resilience theory and neoliberalism. But these paradigms do share profoundly similar epistemological and ontological traits, as Jeremy Walker and Melinda Cooper have shown in an important article.[62] In its own way, resilience theory, with its decisive influence on adaptation to climate change, itself propagates the gospel of flexibility.

At the start of the 1970s, the Canadian biologist Crawford 'Buzz' Holling demolished the idea that ecosystems essentially tend towards stability. As against this vision of nature as an 'equilibrium', he proposed a reading of ecosystems in constant evolution, which could experience multiple states of stability as they rapidly passed from one state to another. Systems do not function in a linear fashion. Or, in other words, the relations between two variables therein are not constantly proportional: knowledge of one particular state of the system does not necessarily give an indication of its future state.[63]

Thus, *constantly* drawing on some resource (for instance, overfishing) does not mean *gradually* exhausting its stocks: rather, this could perfectly well lead to a sudden and unexpected collapse. Resilience is thus both a positive quality of ecosystems and a problem. The resilience which allows for the absorption of small changes without damage may also make classic Malthusian limits invisible. If that is the case, then the users of a resource will not be forced to change their behaviour, and they will continue onward, up to a point when

---

62   Jeremy Walker and Melinda Cooper, 'Genealogies of Resilience: From Systems Ecology to the Political Economy of Crisis Adaptation', *Security Dialogue*, vol. 42, no. 2, 2011, pp. 143–60. The theory of resilience is especially compared to the Austrian trend of neoliberalism (von Hayek, von Mises) rather than to its positivist tendency as represented by the Chicago School (Gary Becker, Milton Friedman), especially in terms of their stress on epistemological limits and complex systems: see Melinda Cooper, 'Complexity Theory After the Financial Crisis', *Journal of Cultural Economy*, vol. 4, no. 4, 2011, pp. 371–85.

63   One useful presentation of Holling's ideas and resilience appears in Raphael Mathevet and François Bousquet, *Résilience et environnement: Penser les changements socioécologiques*, Paris: Buchet Chastel, 2014. One of the best critiques of resilience and adaptive management is that proposed by Paul Nadasdy, 'Adaptive Co-Management and the Gospel of Resilience', in Derek Armitage, Fikret Berkes and Nancy Doubleday (eds), *Adaptive Co-Management*, Vancouver: UBC Press, 2007, pp. 208–26.

this resource abruptly deteriorates.[64] Similarly, the desire to control hazards (for instance, by fighting forest fires) contributes to a build-up of factors (in this case, a stock of combustible brush and wood) that could make some later event catastrophic and precipitate an abrupt change (which small forest fires would have rendered avoidable). The desire to control resources – and the postulate of stability which that desire presupposes – was criticised by Holling, who instead described systems as unpredictable and impossible to plan by their very essence.

Holling then sought to extend his conclusions on the management of ecosystems to much wider domains, and in particular to social institutions. His understanding of social phenomena was informed by his notion of the impossibility of predicting the world and therefore the impossibility of planning. For instance, in one 1971 article, Holling and his co-author, the economist Michael Goldberg, established an analogy between the study of ecological systems and the study of social systems, on the grounds that both were complex systems with analogous structural properties.[65] They provided multiple examples of the unintentional (negative) consequences of attempts at controlling the development of urban space. They considered particularly harmful three policies which were, at that time, largely seen as successes: rent controls to help the poorest; urban renewal programmes; and motorway expansion. In all three cases, the apparent resolution of the problem (high rents, dilapidated buildings, traffic congestion) produced a perverse effect: investors no longer built homes, inhabitants were pushed out, and traffic levels rose. The pair drew a trenchant conclusion:

> We should be much more wary of success than failure . . . Success has given us freeways, urban renewal, and public housing projects. We

---

64   C. S. Holling, 'An Ecologist View of Malthusian Conflict', in Kerstin Lindahl-Kiessling and Hans Landberg (eds), *Population, Economic Development and the Environment*, Oxford: Oxford University Press, 1994, pp. 78–104.

65   Crawford S. Holling and Michael A. Goldberg, 'Ecology and Planning', *American Institute of Planners Journal*, vol. 37, no. 4, pp. 221–30, cf. p. 226.

must reduce the size of our institutions to ensure their flexibility and respect for the system of which they are a small interacting part.[66]

While they named it only roundaboutly ('our institutions') Holling and Goldberg were in truth attacking the state's capacity to intervene in the production of urban space by means of public policy, rather than allowing the market to do its work: the apparent successes of such policies were in fact failures. The state's will to plan and manage suffered from perverse consequences which increased the very problem that it had initially sought to rein in. For Holling and Goldberg, the state should therefore limit its role to incentivising users to make socially desirable choices – for example, by staggering car use in towns over time slots in order to avoid traffic jams. But how could it incentivise them to make such choices? Through prices – that is, through a flexible institution, the market. By setting a price for using roads, and making this price vary as a function of demand, local authorities could modulate residents' behaviour so that they would conform to produce the most effective functioning of the system.[67]

The interesting part of Holling's work – and its importance – is thus its practical aspect, aimed at transforming the management of natural resources. If the quest for stability and equilibrium were illusory or even counter-productive, then approaches to management would have to take this into account. And these approaches would themselves have to evolve and be open to the unexpected, to complexity, to rapid change – in short, to flexibility. Officials could no longer be planners who controlled resources as much as possible; rather, they had to become managers who would learn from their experience, innovate, experiment and, most importantly, be able to adapt quickly to change.

Holling worked with the IIASA (see chapter 1) from 1973 and was its director from 1981 to 1984; it was under its auspices that he

---

66   Ibid., p. 226.

67   Here we have an example of the environmental interventionism which Ferhat Taylan analysed on the basis of the works of Michel Foucault, seeking to modify individuals' behaviour by altering their surroundings. Ferhat Taylan, 'L'Interventionnisme environnemental, une stratégie néolibérale', *Raisons Politiques*, no. 52, 2013, pp. 77–87.

produced one of the most concrete translations of his analyses. Together with his close collaborators, especially William C. Clark (a future big name in sustainability sciences, an IPCC contributor, and today a Harvard professor), Holling postulated the possibility of creating a science of environmental management which would allow a combined treatment of both social and natural systems.

This know-how in 'adaptive management' was meant to allow the managers of natural resources to do their jobs better.[68] What was at stake in this approach was to manage to integrate uncertainty (and thus risks) into decision-making mechanisms which would, at the same time, recognise the existence of ecological limits.

In the volume published in 1978, Holling argued that society's desire to protect itself from the risks created by environmental limits led to the adoption of 'arbitrary, inflexible, and unfocused' regulatory measures.[69] Conversely, while his approach itself sought to reduce uncertainty through the gradual accretion of understanding, it also sought to profit from uncertainty and surprise. To that end, more flexible and adaptable institutions would have to be put in place. Citing the sociologist Michel Crozier, Holling denounced institutions which sought stability as incapable of profiting from change, risk-averse and unable to see crises as opportunities. To them, he opposed the project of:

> learn[ing] to see the world in a new perspective – a perspective that recognizes adaptability and responsiveness rather than prediction and tight control, and a perspective that actively views uncertainty as a fundamental facet of environmental life rather than as a distasteful transition to attainable certainty.[70]

Within the IIASA, William Clark became an important resource in terms of developing Holling's approach and extending it from the

---

68   C. S. Holling (ed.), *Adaptive Environmental Assessment and Management*, IIASA and UNEP, Chichester, UK: John Wiley, 1978.
69   Ibid., p. 6.
70   Ibid., p. 139.

management of local socio-ecological systems (the management of a forest, or of local fish stocks) to problems of global significance like the management of the biosphere itself.[71] Clark rapidly plunged into the climate question, corresponding with other IIASA experts, especially economists, in an attempt to translate climatology's increasing understanding into knowledge that could be used by political decision-makers. He gave the main presentation at the Villach meeting on the climate in 1985 (one of the conferences which led to the Toronto conference in 1988). Widely drawing on Schelling's works, Clark regretted that 'the environmental perspective is inherently biased towards preemptive policies that change the way human activities affect the environment, rather than adaptive towards policies that change the way in which the environment affects human activities.'[72] He then elaborated the array of adaptation possibilities that could be enacted in order to deal with climate change, as well as possibilities for climate geoengineering.

For his part, Holling was less sure that global warming could be overcome so easily. Rather, he believed, the risk was that the resilience of the biosphere would absorb gradual changes without too much damage, before then suddenly swinging into a different state.[73] The changes building up across the whole biosphere on account of human activity had become so enormous and – because they were global – irreversible that adaptation to environmental changes became an inevitability. In particular, the damage to the atmosphere indicated that 'there is now less of a priority to develop predictive tools than to design systems with enough flexibility to allow recovery and renewal in the face of unexpected events – in short, there needs to be a better balance established between anticipation, monitoring, and adaptation.'[74]

---

71   William C. Clark and R. E. Munn (eds), *Sustainable Development of the Biosphere*, Cambridge: IIASA and Cambridge University Press, 1986.

72   William C. Clark, 'On the Practical Implications of the Carbon Dioxide Question', working paper, United Nations Environment Programme, WMO/ICSU/UNEP International Assessment Conference on the Role of Carbon Dioxide, Villach, Austria, 9–15 October 1985, IIASA working paper WP-85-43, 1985, p. 15.

73   C. S. Holling, 'The Resilience of Terrestrial Ecosystems: Local Surprise and Global Change', in Clark and Munn (eds), *Sustainable Development of the Biosphere*, p. 295.

74   Ibid., p. 313.

In reality, the influence of mankind on the environment, and vice versa, would be heightened to such a degree in our era that these systems would co-evolve, at a planetary scale. And these reciprocal effects would now be so powerful that surprises would constantly crop up, that 'challenge traditional human modes of governance and management and that threaten to overwhelm the adaptive and innovative capabilities of people.[75] Consequently, 'the future is not just uncertain; it is inherently unpredictable.'[76]

Faced with this unpredictable world, the first strategy chosen by societies seeking to protect themselves is 'to seek a spurious certitude by increasing control over information and action. The USSR learned the price of that strategy!'[77] Rather less polemically, Holling attacked the vision of sustainable development promoted by the Brundtland report on the grounds that it followed a static, conservative vision of nature and society. In particular, sustainable development and its determination to manage resources would lead to 'more brittle ecosystems, more rigid management institutions and more dependent societies.'[78] Conversely, Holling argued, 'Effective investments in a sustainable biosphere are therefore ones that simultaneously retain and encourage the adaptive capabilities of people, of business enterprises, and of nature. It is the effectiveness of those adaptive capabilities that can turn the same unexpected event (e.g. drought, price change, market shifts) into an opportunity for one system, or a crisis for another.'[79]

Holling drew this sense of constant renewal – and of the capacities for regeneration and reinvention which it permitted – from the Austrian economist Joseph Schumpeter (and, via him, Nietzsche). In

---

75   C. S. Holling, 'An Ecologist View of Malthusian Conflict', in Kerstin Lindahl-Kiessling and Hans Landberg (eds), *Population, Economic Development and the Environment*, Oxford: Oxford University Press, 1994, p. 81.

76   Ibid., p.100.

77   C. S. Holling, 'New Science and New Investments for a Sustainable Biosphere', in AnnMari Jansson, Monica Hammer, Carl Folke and Robert Costanza (eds), *Investing in Natural Capital: The Ecological Economics Approach to Sustainability*, Washington, DC: Island Press, 1994, p. 59.

78   Ibid., p. 64.

79   Ibid., p. 274.

particular, he made use of Schumpeter's theory of 'creative destruction', which sought to understand firms' capacity to innovate in a capitalist economy.[80] In an imperfect world, markets allowed for the most
efficient resource allocation, precisely because they themselves elimi-
nated inefficient businesses. Never did Holling present the state or
democratic institutions as vectors of innovation or as forces which
could stimulate the capacity for adaptation.

CLIMATE, UNCERTAINTY, FLEXIBILITY

This was the context in which the politics of climate uncertainty
would be used by those who, like Lester Lave, denounced the sclerosis
of the existing institutions. Individuals – but also companies, communities, institutions and states – ought to be able to respond on a
continual basis to the signals being sent out by a changing environment and thus to revise their plans to make them compatible with the
latest information. It would thus be illusory to count on stability and
make any long-term predictions.

Adopting Michel Crozier's thesis, Maurice Strong – organiser of
the 1972 summit in Stockholm and a patron of the UNEP – argued
that given the environmental problems faced,

> a drastically new concept of management is vital – one which leads
> in the opposite direction from overblown, centralized, permanent
> super-bureaucracies with decision-making authority concentrated
> at the top. Indeed I believe that we are fast approaching, and in some
> cases may have exceeded, the effective limits of centralized bureau-
> cratic structures, especially in public institutions.[81]

It was thus counter-productive to rely on bureaucratic institutions to handle this new situation; for they were held to be inflexible

---

80   Thus, where Darwin drew a crucial mechanism for his theory of evolution – the
selection of the fittest – from Malthus, Holling found his – creative destruction – in
Schumpeter. Political economy continues to inspire biologists.

81   Maurice Strong, 'One Year After Stockholm: An Ecological Approach to
Management', *Foreign Affairs*, vol. 51, no. 4, 1973, p. 703.

by their very nature. The data which had to be taken into account were so vast in scale, came from such different sources and demanded such particular competences that any centralised body would be incapable of gathering and processing all of them. This debate was of course not new. It had reached its highest theoretical level in the 1930s and 1940s, when socialist theorists of democratic planning like Otto Neurath, Oskar Lange and Karl Polanyi debated the apologists of economic liberalism like Ludwig von Mises and Friedrich von Hayek[82] – who went on to inspire Reagan and Thatcher.

For Hayek, there already existed a social mechanism – the market – whose operation supposedly ensured a synthesis of information better than any centralised mechanism could achieve. By allowing independent agents to react spontaneously to changes in their environment, by translating their needs and demands into a universal language – prices – the market could theoretically enable constant and quasi-spontaneous adaptation to all change. And companies, rather than states, were the most reactive entities. Already during the preparatory work for the Stockholm conference, Maurice Strong had advanced the consideration that 'the institutions of industry are even more adaptable than the institutions of governments, and this adaptability is one of the greatest assets that they bring to play'.[83] Nor did this quality escape Thomas Schelling, who argued:

> With full recognition that the market cannot respond to all environmental or foreign policy considerations, it remains true that for most business and consumer decisions the market, as a process, has the important virtues of flexibility and adaptability.[84]

The political scientist Dean Mann, a participant at the Annapolis seminar in 1978, asserted that responses to the climate question risked

---

82 John O'Neill, 'Knowledge, Planning, and Markets: A Missing Chapter in the Socialist Calculation Debates', *Economics and Philosophy*, vol. 22, no. 1, 2006, pp. 55–78.

83 'Use Now, Pay Later?', interview with Maurice Strong, in *Alfa-Laval International*, 1970–1971, p. 13.

84 Schelling, *Thinking Through the Energy Problem*, p. 62.

'overloading' the political system, in particular when it came to attempts to reduce CO_2 emissions. But it was possible to lower this risk by choosing a strategy of *adaptation* rather than *reduction*, which would allow 'private decisions under market-like conditions to achieve societal goals. In other words, both rules and prices might be allowed to play roles in the achievement of societal adjustment to climate change.'[85]

In the 1970s and 1980s, the attack on rigid state institutions, coupled with a more or less open apologia for a market-based *modus operandi* – all in the name of dealing with environmental constraints – constituted a breeding ground for neoliberalism. And the climate question held an important place in this regard. The qualities of flexibility, adaptation and resilience would be used to respond to the challenges posed by climate change.

Over the course of the 1980s, the AAAS working group on the climate continued its work. After first taking interest in food security and agricultural research, it delved into the implications of climate change for water management in the United States. In 1987, its director, David Burns, sought to establish a five-year development plan for its activities. In particular, he suggested

> a project to identify strategies which increase flexibility and improve the ability of a country (or region, or economic sector) to exploit favorable climate change, and adapt or cope most easily to unfavorable change. The project would aim to identify actions which are needed and are cost-effective now, but which could be vital in a period of rapid climate change.[86]

In response to this suggestion, Thomas Schelling replied that 'this is the kind of thing I am always in favour of'.[87] Jesse Ausubel suggested

---

85   Dean E. Mann, 'Research on Political Institutions and Their Responses to the Problem of Increasing CO_2 in the Atmosphere', in Chen, Boulding and Schneider (eds), *Social Science Research and Climate Change*, p. 126.

86   David Burns, Memorandum, 'Possible Future Activities – AAAS Climate Project', 5 January 1987, AAAS, Washington, DC, AAAS Archives, box 15.

87   Letter from Thomas Schelling to David Burns, 10 November 1986, AAAS Archives, box 15.

reflection on 'mechanisms to involve many people directly in thinking about how they might change their behavior':[88] rather than seeking government action, he proposed that this thinking be delegated to polluting companies themselves. He added, 'significant progress is more likely to come from a bottom-up approach, where many people and firms and organizations are making widespread, voluntary, positive choices to prevent, adapt, etc., than from a top-down approach.'[89]

For US experts, at the confluence between science, senior administration and the think-tanks working on development and resource management policy, the gospel of flexibility – and thus the extension of the market's domain – became a self-evident truth. Indeed, it was an assumption embedded within their work in a great variety of contexts.

The Resources for the Future foundation very early set to work formulating economic solutions for environmental problems – solutions which respected the canons of neoclassical economics.[90] With a view to preparing the position the US would take within the recently established IPCC, RFF organised a seminar with funding from the main US agencies concerned. Held in Washington in June 1988, it was titled 'Controlling and adapting to climate warming'. Its director, Paul Portney, quite simply asserted that the refusal to accept climate change and its beneficial consequences ('a gentler climate', as he saw it) expressed a general attitude of inflexibility. Such an attitude consisted of refusing all change, or what he called 'inherent biases we feel in favor of the status quo'.[91] As for the economist Pierre R. Crosson, another RFF researcher, he turned Heilbroner's argument inside-out, noting that 'countries whose institutional structures permit a ready shift of resources among uses in response to changing conditions of

---

88   Jesse Ausubel, Memorandum (to David Burns), 3 November 1986, AAAS, Washington, DC, AAAS Archives, box 15, p. 1.

89   Ibid.

90   Bonneuil and Mahrane, 'Gouverner la biosphère', pp. 137 and 155.

91   Paul Portney, 'Assessing and Managing the Risks of Climate Change', in Norman Rosenberg, William E. Easterling III, Pierre R. Crosson and Joel Darmstadter (eds), *Greenhouse Warming Abatement and Adaptation*, Washington, DC: Resources for the Future, 1989, p. 87.

scarcity and surplus should have greater adjustment ability than countries with less flexible institutions.'[92]

CASHFLOW . . . FROM WATER

If we want to trace how the gospel of flexibility spread as a response to climate change, water provides a particularly revealing case study. Access to drinking water is necessary for life itself. We might, therefore, think that it is a common good – and that it would be an obvious condition of any social contract that authorities should ensure that enough of it is supplied to each inhabitant. The cost of this supply would then be covered by taxation or an annual licence. The population's capacity, or rather, incapacity, to pay ought not to restrict its access to this fundamental good. Such is the conception of water supply as a public service, established in the industrialised countries in the late nineteenth and early twentieth centuries, especially thanks to the actions of municipal authorities. Even if the actual provision of this resource is outsourced to some private company, drinking water is not a commodity whose cost should be determined by the play of supply and demand: the tariff should continue to be regulated by the public authorities. In fact, having experienced repeated operational failures and endemic corruption on the part of private water management companies, many jurisdictions which had outsourced or privatised their water distribution systems are now seeking to 'remunicipalise' it, taking it back into public hands. Citizen social movements like that in Cochabamba, Bolivia, in 2000, or more recently in Ireland, have waged 'water wars' to oppose the commodification of this essential public good.[93]

For the neoliberals, conversely, the inability to turn water into a simple commodity is problematic, for it stops the free market from doing its work of allocating resources (and, incidentally,

---

92 Pierre R. Crosson, 'Climate Change: Problems of Limits and Policy Responses', in ibid., p. 74.

93 Daniel Finn, 'Water Wars in Ireland', New Left Review, no. 95, 2015, pp. 49–63.

prevents water multinationals from making profits). Yet since the early 1980s certain economists and political scientists have invoked adaptation to climate change as an argument that justified a policy favouring the increased commodification of water. At the Annapolis seminar, for example, Dean Mann opined that the constraints imposed by climate change would necessitate 'the creation of the water markets in which water is applied to its highest economic use',[94] in order to arbitrate between the various sources of demand on this resource.

RFF's Kenneth Frederick and Peter Gleick advocated that market mechanisms be introduced into water management in order to respond to the imperative of climate adaptation:

> Improved management of existing supplies, marginal-cost pricing and water marketing to encourage conservation and reallocation of supplies in response to changing conditions, and development of water-saving technologies can provide considerable flexibility for adaptation.[95]

They launched a full-on attack against non-commodifiable (and thus non-transferable) water usage rights, public controls, subsidies and the regulation of water tariffs. In their view, by limiting market action all these things combined to reduce the water sector's capacity to adapt to climate change, and therefore to increase its impact.

When in the early 1990s the US National Academy of Sciences produced its 'Synthesis Report' on the policy implications of climate change, its panel of experts included Stephen Schneider, Crispin Tickell and Maurice Strong, and the economists William Nordhaus and Robert Solow. They concluded that in addition to the development of agronomic research, the main adaptation measure that needed to be taken was to 'make water supply more robust by

94 Mann, 'Research on Political Institutions and Their Responses', p. 135.

95 Kenneth D. Frederick and David C. Major, 'Climate Change and Water Resources', *Climatic Change*, vol. 37, 1997, pp. 7–23.

coping with present variability by increasing efficiency of use through water markets and by better management of present systems of supply.'[96]

These solutions were also encouraged by the first IPCC experts' report, completed in 1990. It, too, advocated a more flexible management of water supplies based on the price mechanism. This meant the creation of water markets (demanding that clear property rights be stipulated) and, more generally, pricing according to the estimated true costs of water – which, it was supposed, would reduce wasteful use of this resource.[97] The concrete measures proposed by the IPCC in the section devoted to 'resource use and management' had to do above all with water use and agriculture. These measures most importantly consisted of intensifying agricultural production by boosting efficiency and productivity. The requirements of adaptation to climate change were the justification for the commercial orientation of agriculture – on the productivist model inherited from the 'green revolution' – in response to rising demand for agricultural produce. Even if, in formal terms, the countries of the Global South did contribute to the preparation of these reports, it is striking to note that the US took the presidency of working group three, which dealt with the 'response strategies' for the climate question. The Assistant Secretary of State Frederick Bernthal, a Reagan appointee, used the ample resources of the US administration to compile this report.[98] When it came to the section on the use and management of resources, for example, the US obtained no fewer than nine delegates on the relevant commission (and their British allies four), while most of the other participating countries (China, France, Brazil, India, USSR, Nepal, Switzerland, Zimbabwe) had

96   National Academy of Sciences, *Policy Implications of Greenhouse Warming Synthesis Panel*, Washington, DC: National Academy Press, 1991, p. 77.

97   IPCC, Climate Change, *The IPCC Response Strategies*, WMO/UNEP, 1990, pp. 180–3.

98   See Howe, *Behind the Curve*, pp. 163–5. However, I disagree with Howe's contention that working group three produced only 'technical solutions' to climate change. Rather, it proposed an array of solutions, based on the political economy of flexibility.

only one or two. In addition, the five 'civil society' observers counted two RFF delegates, one from the World Resource Institute, and a lobbyist for the coal industry![99]

With all this intense work to spread the gospel of flexibility, it is hardly surprising that when, in 1990, the IPCC first elaborated its thinking on the substance of the adaptive policies that should be adopted in response to the challenge of climate change, the first criterion it stipulated was . . .

> *Flexibility.* Since the effects of climate change are uncertain, responses need to be successful under a variety of conditions, including no-climate-change. Thus, flexibility is a matter of keeping options open. For instance, a market mechanism for pricing and allocating resources will work under a variety of conditions and, therefore, is flexible.[100]

FORGETTING ADAPTATION

If, in the US context, the response to the impacts of global warming consisted of an appeal to the gospel of flexibility, this would briefly be oriented – for just over a decade – towards an international reduction of greenhouse gas emissions. The Toronto conference, the establishment of the IPCC and the negotiation of the United Nations Framework Convention on Climate Change (UNFCCC) in Rio in 1992 would essentially concentrate on reduction policies. This began a period (especially marked by the Kyoto Protocol) which would last up till the COP summit in Copenhagen in 2009, which marked the end of international will to reduce greenhouse gas emissions. During these years, there was of course much talk of flexibility, elbow room and markets, but, above all, of the need to soften the effects that climate regulation would have on businesses.

Contrary to one commonplace reading, adaptation did not 'emerge' in the early 2000s as a palliative in response to the failures of

---

99   For a list of the delegates, see IPCC, *Climate Change*, pp. 203–5.
100   Ibid., p. 175.

reduction policies. Rather, adaptation – and its gospel of flexibility – had made headway, underground, during the very years in which international climate policy appeared to sweep all before it. And when adaptation did re-emerge, it donned the same clothes that had been tailored for it in the 1970s and 1980s by a combination of researchers, scientific bodies, private foundations and government agencies. These had each converged on a common definition of the climate question – as a politics of uncertainty – and on the formulation of the solution in the gospel of flexibility. The goal pursued by the champions of such an adaptation was less a matter of pressing on with fossil fuel extraction (though this was obviously a motivation for some of them) as of the promotion of a radically different vision of society, very much in step with the neoliberal counter-revolution and the extension of the market.

Chapter 3
Climate and Market: The Double Shock

Naomi Klein has highlighted how neoliberal ideologues and politicians can exploit the 'shock' of some disaster in order to impose a market 'shock' on a disoriented society.[1] Policies that would be unacceptable in normal times, when they are subject to public deliberation and democratic debate, may in times of emergency appear to be the only possible response. In the same vein, Jamie Peck speaks of 'fast policy transfer',[2] advancing on ground already prepared by a strategy of 'permanent persuasion' on the ideas front.

The rebuilding of New Orleans after much of it was destroyed by Hurricane Katrina in 2005 thus served as a sandbox for a series of neoliberal and racist policies. A majority Black and poor city in the South, New Orleans has long been the target of conservative attacks stigmatising a Black, welfare-dependent population whose mere presence was blamed for middle-class white flight . . .

Less than two weeks after the Katrina disaster, a neoliberal think-tank, the Heritage Foundation, published a 'ready to roll out' programme for rebuilding the city. This programme demanded deregulation in order to drive 'private investment', the implementation of federal tax cuts to create a favourable climate for innovation and the distribution of aid in the form of vouchers to give families 'freedom of choice' in their spending. The emergency finance freed up

---

1 Naomi Klein, *The Shock Doctrine: the Rise of Disaster Capitalism*, London: Penguin, 2014.

2 Jamie Peck, *Constructions of Neoliberal Reason*, Oxford: Oxford University Press, 2010, pp. 137ff.

in order to assist the inhabitants of New Orleans would have to be deducted from existing federal funds. This meant abandoning other federal spending.[3] In the weeks and months following the disaster, New Orleans was subject to military occupation. Formally, this was a matter of avoiding 'looting'. But above all, it was aimed at imposing order among the Black population, by violence if need be.

The Heritage Foundation's proposals were listened to attentively at all levels of government. The 'rebuilding' of the city was founded on three principles: first, on the idea that this was a local problem and that federal interventions did more harm than good; second, that the costs of rebuilding must not be added on to existing spending; and third, that the population ought to rely on itself to get out of this situation and not 'expect the state to do everything'.[4] These principles of course aggravated existing structural injustices in the city, which, ten years later, proved even more unequal than before the disaster.

With a certain dose of gallows humour, a few months after Katrina, Professor Robert Bullard, a founding father of studies on environmental justice, proposed a twenty-point reconstruction plan 'to destroy black New Orleans'. In particular, he suggested that the city's Black-majority low-lying districts should not be rebuilt (in order to 'protect the swamp') and to concentrate depollution efforts on white and wealthy districts. In fact, many long-standing Black districts were destroyed exactly for these reasons, allowing private developers fresh ground to accommodate new, high-income residents. Ten years later, Bullard was forced to acknowledge that his 'plan' had been followed to the hilt and that the city now counted far fewer African Americans, with increased inequality, a spike in urban poverty among Black children, a shortage in social housing, a sharp rise in rents, continued discrimination in housing, massive gentrification of certain neighbourhoods and a generally unequal reconstruction.[5]

---

3   Ibid, pp. 152–8.
4   Ibid., p. 176.
5   Robert Bullard, 'Katrina and the Second Disaster: A 20-Point Plan to Destroy Black New Orleans Revisited After 10 Years', blog, 28 July 2015, drrobertbullard.com.

Disaster neoliberalism implies a reorganisation of the state (less public transport, less subsidised housing, fewer public hospitals, more police and prisons, etc.). But it also means an attempt to reorganise individual and collective behaviours by weakening those social relations which do not yet depend on the market. Insisting on individual responsibility, private initiative and, more generally, on the inefficiency of the state (except in terms of maintaining order), the neoliberals propose a vision of freedom reduced to the freedom of trade and commerce.

But this freedom does not emerge spontaneously. Rather, it has to be organised. This is the great lesson of the German ordoliberalism of the 1930s (Walter Eucken, Franz Böhm, Alexander Rüstow, Wilhelm Röpke, Alfred Müller-Armack), a major source of contemporary neoliberal thought.[6] Just like the commons of the medieval era, employment and income security, predictability and rest time are working-class institutions which need eradicating in any modern society.

For these theorists, a market economy must be based on entrepreneurial individuals who, unable to count on social security, are thus driven to constantly outdo themselves, to innovate, to train, to act on a continual basis. One's security must be an individual matter, based on small property ownership (for instance, housing, or building up capital for one's retirement) rather than in any collective way. These individuals thus become capable of surviving external shocks, independently of any state support. The theorist Werner Bonefeld shows how, for ordoliberals, this 'vital policy' has to be actively produced by the state – and if necessary, an authoritarian state – failing which, the risk is that a 'proletarian policy' of social security will instead prevail.[7]

In this chapter, we will see how neoliberal ideas infuse contemporary responses to the challenge of adapting to climate change. In this

6   Werner Bonefeld, 'Human Economy and Social Policy: On Ordoliberalism and Political Authority', *History of the Human Sciences*, vol. 26, no. 2, 2013, pp. 106–25.

7   Werner Bonefeld, 'Freedom and the Strong State: On German Ordoliberalism', *New Political Economy*, vol. 17, no. 5, 2012, pp. 633–56.

domain, capitalism enjoys such 'ecological dominance'[8] that most of the solutions envisaged, enlisted and enacted by the big international institutions themselves contribute to the wider neoliberal realignment: that is, by 'neoliberalising adaptation'.[9] To put it bluntly, the shock of global warming is being used to extend market mechanisms and to increase the integration of marginal populations into the global market. Adaptation to climate change is thus part of the extension of the market and – on that basis – of the primitive accumulation of capital.

MICROFINANCE: AN EXTENSION OF THE DOMAIN OF ADAPTATION

Climate change, and most importantly the rise in climate variability, have become pretexts for the extension of economic rationality into spaces where it had not previously dominated. One striking example of this is apparent in the numerous 'microinsurance' projects developed by international organisations, private companies and charities in order to allow poor peasants in the Global South to adapt to climate change. Microinsurance is, together with microcredit and microsavings, one of the three main branches of microfinance. The rise of the prefix 'micro' in development policies is testament to a rather telling change of scale. After the fall of the USSR, any attempt at global social transformation has been abandoned in favour of 'micro-' reforms and institutional adjustments.[10]

The theory, and indeed the practice, of development has largely been reoriented to the idea that the only acceptable economic model is a capitalist one. In practical terms, it thus becomes necessary to create

8   Bob Jessop, 'The Crisis of the National Spatio-Temporal Fix and the Tendential Ecological Dominance of Globalizing Capitalism', *International Journal of Urban and Regional Research*, vol. 24, no. 2, 2000, pp. 323–60.

9   Romain Felli and Noel Castree, 'Neoliberalising Adaptation to Environmental Change: Foresight or Foreclosure?', *Environment and Planning A*, vol. 44, no. 1, 2012, pp. 1–4.

10   Isabelle Guérin, *La Microfinance et ses dérives: Émanciper, discipliner ou exploiter?*, Paris: Demopolis, 2015, provides a critical analysis of microfinance and microcredit.

the institutional and cultural conditions that would force individual choices and public policies to be oriented toward this type of accumulation.[11] After inflicting austerity treatment (so-called 'structural adjustment plans') on the countries of the Global South in the 1980s according to a crudely neoliberal vision ('less state!'), the World Bank and the International Monetary Fund revised their approach to development somewhat towards the end of the 1990s and in the early 2000s. The retreat of the state and the public sector was no longer an end in itself. Rather, it was designed to facilitate the development of an efficient, flexible, extended market economy which would enjoy a certain legitimacy among the population.[12] In a certain sense, these organisations abandoned the idea of a spontaneous market-based order that would emerge when obstacles were removed from its path. They drew rather more inspiration from the ordoliberal approach, which insisted on the need to construct the market order and regulate it to ensure its proper functioning. This was no longer only a matter of creating new markets, but of subjecting social and political institutions to economic rationality.[13] Constructing this new order would require an entrepreneurial spirit, which would have to be awakened among the population in order to counter its 'proletarian' tendencies: each individual was summoned to become an 'entrepreneur of the self'. No longer were there market failures, just individuals who had not yet conformed to the demands of the market sufficiently. Following in the tracks of Michel Foucault, many analyses now spoke of a neoliberal 'governmentality', to designate these attempts at reforming behaviours.

The recent success – in international organisations as in public debate – of Esther Duflo's behavioural economics of development is

---

11   Colin Leys, 'The Rise and Fall of Development Theory', in Colin Leys, *Total Capitalism: Market Politics, Market State*, Monmouth, UK: Merlin Press, 2008, pp. 9–64. For an analysis of the most recent turns in development economics, see Bengi Akbulut, Fikret Adaman and Yahya M. Madra, 'The Decimation and Displacement of Development Economics', *Development and Change*, vol. 46, no. 4, 2015, pp. 733–61.

12   Jamie Peck and Adam Tickell, 'Neoliberalizing Space', *Antipode*, vol. 34, no. 3, 2002, pp. 380–404.

13   Akbulut, Adaman and Madra, 'Decimation and Displacement of Development Economics', p. 747ff.

symptomatic of this change. She proposes policies which are 'pro-poor' but are based on the market and dependent on the reorientation of their economic behaviour.[14] In this approach, poverty is explained not by the economic dependence of the countries of the Global South, the unequal exchange that dominates global trade, tax evasion by the dominant classes, the stealing of land and the pillage of natural resources, exploitation, oppression and marginalisation. The idea that the poverty of the mass of people results from the accumulation of wealth, and thus from growth itself, is absolutely alien to this analysis. On the contrary, here everything is the result of individual choices and the domestic policies of the Global South countries themselves. Hence, these choices and policies need to be 'pragmatically' reformed, with the help of econometrics, in order to 'nudge' economic behaviours, following the 'liberal-paternalist' programme of the economist Cass Sunstein.[15] The fight against poverty would then be advanced through the reduction in the number of schoolchildren per class, or by making mosquito nets available for free, rather than through land reform or nationalisations.

The World Bank seeks to remodel the state and existing institutions so that they will encourage a favourable attitude towards capitalist accumulation: with security for property rights, a workforce dependent on market-sourced income, freedom of capital investment and a flexible labour market. Paradoxically, this requires a certain paternalist benevolence, embodied in minimal forms of social security (a 'social safety net'), poverty reduction targets and public investment in order to develop 'human capital' (through education,

---

14   See Abhijit V. Banerjee and Esther Duflo, *Poor Economics: A Radical Rethinking of the Way to Fight Global Poverty*, New York: Public Affairs, 2011. For a detailed critique, see Sanjay G. Reddy, 'Randomise This!: On Poor Economics', *Review of Agrarian Studies*, vol. 2, no. 2, 2012, pp. 60–73. See also Christian Berndt, 'Behavioural Economics, Experimentalism and the Marketization of Development', *Economy and Society*, vol. 44, no. 4, 2015, pp. 567–91. According to the *Nouvel Observateur* (5 November 2015), Duflo is one of the '15 left-wing intellectuals worth knowing'.

15   John Cassidy, 'Economics: Which Way for Obama?', *New York Review of Books*, 12 June 2008; Jamie Peck, *Constructions of Neoliberal Reasons*, chapter 6.

professional training, healthcare, etc.).[16] Capital wants a sufficiently numerous, skilled and educated workforce to be available, without itself having to bear the costs of its reproduction. These costs are socialised and certain minimal foundations of social reproduction are provided by the state. But rather than base themselves on a welfare state which extends universal free public services to the population as a whole – as in a Polanyian counter-movement to protect society from the market – these new forms of 'social protection' have promoted populations' maximum individual dependence on the market. To adopt the economist Susan Soederberg's analysis, they 'provid[e] a market-led solution to the social reproduction of the workforce'.[17] The same goes for the – today very fashionable – conditional cash transfer programmes developing in Latin America, in which recipients receive small amounts of money on condition that they follow certain guidelines (for instance, having their children vaccinated or sending them to school).[18] The notion of 'human development', promoted by the United Nations Development Programme (UNDP) is the humanitarian twin of this constructivist reorientation of international development policies.

MICROCREDIT: TO ADOPT IT IS TO ADAPT

The promotion of microcredit is part of this reorientation. In the 2000s, it made it possible to believe that an effective solution to poverty in the countries of the Global South had finally arrived. If society is 'naturally entrepreneurial', all that was needed was to provide some starter capital to the masses, who were only waiting for this small stepladder in order to launch a productive enterprise. By multiplying their loans of very small amounts of money – by way of

16 Marcus Taylor, 'Responding to Neoliberalism in Crisis: Discipline and Empowerment in the World Bank's New Development Agenda', *Research in Political Economy*, vol. 21, 2004, pp. 3–30.

17 Susan Soederberg, *Debtfare States and the Poverty Industry: Money, Discipline and the Surplus Population*, London: Routledge, 2014, p. 194.

18 Lena Lavinas, '21st Century Welfare', *New Left Review*, no. 84, 2013, pp. 5–40.

priority granted to women and sometimes guaranteed by collective as a community responsibility – some banks managed to create markets and make a profit, thanks to poverty.[19] In 2006, Muhammad Yunus, founder of the Grameen Bank, received the Nobel Prize for this endeavour.

Enthused by this extension of the market, which seemed to offer the poorest, from India to Africa and Latin America, an entrepreneurial way out of their condition, many international organisations and NGOs – backed up by a swelling academic industry – happily sang the praises of microcredit. The entrepreneurial path out of poverty was especially well-adapted to the fiscal crisis situation in the states of the Global South. For these latter are incapable of financing public services and carrying out a social redistribution of income, even when they want to (notably on account of the tax evasion organised from Northern countries, like Switzerland).

The fight against poverty ceases to be a humanitarian action or a civic task, instead becoming a profitable economic activity. The family resemblance between this thinking and recent metamorphoses in environmental protection is obvious. Both are reduced to their contribution to the production of economic value, and they no longer constitute values (whether moral, ethical or political) worthy of interest on their own account.[20]

Of course, rather than speaking of an extension of capitalism or of the market, international organisations prefer to reach for a euphemism – and talk of 'financial inclusion'.[21] If the poor are poor, this is not because they are subject to the implacable discipline of the market and its laws, but rather because they are 'excluded' from it. 'Financial inclusion' is thus presented as good fortune, a matter of increasing one's 'opportunities' to be an entrepreneur and have one's work pay off in the economic sphere.

---

19   Soederberg, *Debtfare States and the Poverty Industry*.

20   A development noted, among others, by the philosopher Michael Sandel in *What Money Can't Buy: The Moral Limits of Markets*, New York: Farrar, Straus and Giroux, 2012.

21   Soederberg, *Debtfare States and the Poverty Industry*, chapter 7.

But the reality of microcredit is far different from the way it is advertised. The loans which are supposed to allow rural populations to invest in modernising farming or in launching new businesses are rather more often used to meet immediate needs: that is, to buy food, energy or medication. Beyond a few real successes (obligingly trumpeted by media) microcredits have above all served to spread out over space and time the permanent deprivation which the greater part of humanity has to confront. In short, they at first provided a (very) minimal social benefit, but did not serve as a trampoline to entrepreneurship.[22]

All this would not be such a grave problem, if microcredit was charity. Sadly, as its promoters themselves constantly remind us, it is in fact a business. The money thus transferred is not an unconditional basic income, but well and truly a loan, which has to be paid back. The millennia-old business of usury, which consists of lending money to the poorest at prohibitive interest rates, is enjoying a fresh spring of youth. Once the loan has been eaten up, the creditors – or their hired muscle – come along to demand it is repaid. The repayment demands which cannot be honoured drive the poorest peasants to take on further debt at even more unfavourable interest rates, and to concentrate all their activity on the production yielding the highest profits in order to respond to their creditors' demands. Reaching the outer limit of their economic capacities, some – including those who have lent money to a relative or an acquaintance out of solidarity – have to sell their meagre capital or their land. For the poorest populations, the crisis of poverty is thus aggravated by the crisis of indebtedness, as they are caught in a poverty trap not despite but precisely because of microcredit.

---

22   Milford Bateman and Ha-Joon Chang, 'Microfinance and the Illusion of Development: From Hubris to Nemesis in Thirty Years', *World Economic Review*, vol. 1, no. 1, 2012, pp. 13–36.

DROUGHTS DON'T OFFER CREDIT

Nowhere has the crisis of indebtedness-through-microcredit proven quite as tragic as in the Indian province of Andhra Pradesh. As the development specialist Marcus Taylor has shown, this is a crisis intensified multiple times over by the effects of climate change.[23] A rural region with semi-arid zones, Andhra Pradesh was suffering a major agrarian crisis to which the growth of microcredit offered an opportune response. Originally set up by NGOs, and aimed at essentially humanitarian objectives, in the 1980s microcredit was promoted by the World Bank and the Indian state as an entrepreneurial solution to the chronic problems of poverty among marginal populations. The initial humanitarian endeavour was followed by the rise of a private economic sector made up of profit-seeking microcredit institutions. Andhra Pradesh thus became one of the main destination regions for global financial flows seeking to create value through the microcredit sector, though at the beginning of the 2010s this sector amounted to only a few billion dollars.[24]

Yet far from easing the agrarian crisis in Andhra Pradesh, the penetration of microcredit into the region amplified it, pushing hundreds of thousands of small peasants into bankruptcy when they were unable to repay their debts. This crisis assumed a violent form: it led to an unprecedented suicide wave among ruined peasants, almost 25,000 of whom have killed themselves over the last decade and a half.[25] The state was forced to intervene, temporarily suspending the repayment of the debts, which in turn precipitated a crisis in the financial sector.

A complex combination of ecological relations, debt and power moulded the way in which this crisis played out and determined who

---

23 Marcus Taylor, *The Political Ecology of Climate Change Adaptation: Livelihoods, Agrarian Change and the Conflicts of Development*, London: Routledge, 2015, chapter 7.

24 Marcus Taylor, '"Freedom from Poverty Is Not for Free": Rural Development and the Microfinance Crisis in Andhra Pradesh, India', *Journal of Agrarian Change*, vol. 11, no. 4, 2011, pp. 484–504.

25 Taylor, *Political Ecology of Climate Change Adaptation*, p. 144.

the victims would ultimately be. I cannot here provide a detailed account of disaster (for which, see Taylor's '"Freedom from Poverty Is Not for Free"' and *Political Economy of Climate Change Adaptation*), but we can at least dwell for a moment on one of the ecological dimensions of the problem. The need for poor peasants to commercialise and valorise their production as much as possible in order to repay their debts drove them to try to increase their productivity. They were forced to turn away from agriculture producing food for the local population to more economically profitable kinds of output; yet this also demanded major investment, while returns on that investment would depend on the global market. The intensification of production pushed the agricultural model to its ecological limits. Water scarcity was particularly acute in this semi-arid region. The community forms of managing this resource (with community irrigation systems, collective cisterns, etc.) were replaced by an entrepreneurial strategy of well drilling.

This scramble for water led to ecological disaster. The proliferation of drilling exhausted aquifers, and it also required phenomenal quantities of energy to keep the pumps working. But this was also a social disaster: for dispossessed of the best land and deprived of resources, the poorest peasants were incapable of investing the amounts needed for this activity. They then turned to microcredit – but most of the wells dug did not provide water (on account of the generalised rush to use the aquifers) and those who had drilled them were ruined. A study on suicides in the region indicates that in 30 per cent of cases, it was the inability to repay debts owing to well drilling that precipitated the fatal act.[26]

The relations of indebtedness which afflicted the poor peasants were not, therefore, a consequence of bad harvests. The opposite is true: indebtedness is inherent to all production in this system and it functions as an essential (unequal) social relation. The great vulnerability of poor peasants in Andhra Pradesh to climatic variations stands at the end of a chain of dispossession and marginalisation caused by relations of indebtedness and land dispossession, of which

---

26   Ibid., p. 159.

they are the victims. This dispossession is carried out to the profit of the big landowners and financial intermediaries. In other words, the poor peasants' vulnerability is the result of a relationship of class and of capital formation.

Yet the plans for adapting to climate change, currently under discussion in this Indian province, completely overlook the question of the social vulnerability produced by these power relations, which especially materialise in relations of indebtedness. On the contrary, they make adaptation to climate change into an essentially technical question – as if it was all to do with the choice of crops, weather forecasting systems or, at best, the community management of water. In not dealing with the origins of vulnerability – and in particular the agrarian crisis prompted by microcredit-induced indebtedness – this approach, Taylor emphasises, depoliticises the question and validates or even strengthens existing power relations.

OVERCOMING CLIMATE CHANGE WITH INSURANCE

The tarnished image of microcredit has not, however, made any dent in the reputation that microinsurance today enjoys. On the same model as its predecessors, it relies on the principle that tiny streams make up great rivers, and even if the poor do not have a lot in the way of possessions, they do have the advantage of their weight of numbers.

Microinsurance owes its rise to belief in the gospel of adaptation to global warming. Why should poor peasants be particularly vulnerable to the effects of global warming? The problem is presented by insurers as a matter of risks and how they are managed. Since they are resource-deprived, poor peasants do not have a margin of security that would allow them easily to survive bad harvests. One bad season, one insufficient harvest, and they will have to sell their reserves, their tools or their livestock. And selling these things will deprive them of the means by which to raise crops in the season that follows.

Yet global warming has the effect of increasing climate variability – in other words, of creating more uncertainty and thus more risks. Temperature fluctuations and precipitation increases make it more

difficult to forecast the meteorological conditions in which production will take place. Here we again come across the theme of the growing uncertainty of the world and the ontology of risk, which charac-
terises the theories of adaptation and resilience. Their goal, we should
remember, is not to push back against the sources of global warming,
but rather to mitigate its consequences, in a context posed as essentially impossible to know or to plan for.

Insurance multinationals and their counterparts in reinsurance
(insurers for insurers) are among the most active capitalist industries
on the terrain of global warming.[27] And in recent years they have
turned toward adaptation. The world's largest reinsurance firm,
Munich Re, finances research programmes, research grants, reports
and conferences concerning the management of climate risks. In 2005
it founded the Munich Climate Insurance Initiative (MCII) within
the United Nations University (UNU) to serve as the ideological
spearhead for promoting an entrepreneurial vision of adaptation that
would prepare a large role for the private insurance sector. Particularly
active in global climate governance circles, they intervene at the COP
and other sites of international climate policy and expertise.

Indeed, at the Annapolis seminar in 1978, Dean Mann had well
understood that 'in the context of an adaptive strategy . . . an approach
that would emphasize the private sector would be the inclusion of the
insurance industry as a major actor in dealing with climate change.
Either with or without government involvement, perhaps through
some reinsurance arrangement, the private insurance industry might
participate by insuring against the potential effects of the consequences of climate changes'.[28]

Insurers and their political and scholarly outriders today present
insurance as a solution that makes it possible to mitigate the structural uncertainty driven by global warming. By mutualising the risks,
insurance makes it possible to 'spread out' its effects among the

---

27   Razmig Keucheyan, *Nature Is a Battlefield*, Cambridge: Polity Press, 2016, deals
in greater detail with the array of links between insurance and climate risks that extend
beyond the field of adaptation policies and microinsurance: see his chapter 2.

28   Mann, 'Research on Political Institutions and Their Responses', pp. 139–40.

population. Better, in making premiums and compensation dependent on its buyers' own prudential behaviour, insurance polices this behaviour, developing incentives to act according to a strict economic rationality (by saving, by taking calculated risks, by not exposing oneself too much to some danger, etc.). By providing a remedy for the lack of financial resources and the presumed lack of expertise in the countries of the Global South, insurance multinationals present themselves as benevolent agents of adaptation. And they do so with the active collaboration of international organisations.

Thus, the United Nations Environment Programme (UNEP) has for several years been working with the biggest global insurance and reinsurance corporations, especially in the context of its 'Principles for Sustainable Insurance' initiative (PSI). Launched in 2012 in partnership with firms like Axa, Allianz, HSBC and Munich Re, the PSI promotes these private companies' expertise in terms of managing disastrous risks. Dressed up as the pinacle of virtue, the insurance industry is being promoted as the main specialist in risk management. Its role is not limited to compensating for the losses suffered when natural disasters hit, but beyond that to manage the whole chain of risk, from prevention to reconstruction, through management of risk reduction. States, international organisations and civil society are called on to collaborate with these experts and put their knowledge to use: 'The vision of the PSI Initiative is of a risk-aware world, where the insurance industry is trusted and plays its full role in enabling a healthy, safe, resilient and sustainable society'.[29]

In reality, the private insurance sector is far from being as efficient as insurers like to claim. In the industrialised countries, which have a much more developed insurance sector, but also social insurance and strong state regulation, the private sector often plays only a subsidiary role and has limited influence over risk management.[30]

---

29   UNEP Finance Initiative, *The PSI Global Resilience Project: Building Disaster-Resilient Communities and Economies*, Geneva: UNEP, 2014, p. 5.

30   Kristian Krieger and David Demeritt, 'Limits of Insurance as Risk Governance: Market Failures and Disaster Politics in German and British Private Flood Insurance', Discussion Paper no. 80, Centre for Analysis of Risk and Regulation, LSE, October 2015.

Here the insurers have long sought to extend their activities and profits by organising a form of 'solidarity without the state',[31] even if this
solidarity is evidently commercial in nature. And in this regard, they
have partly succeeded! But the principles that undergird the private
insurance sector clash with the need to reduce social vulnerability. In
allowing compensated customers to get back to normal after a disaster – for example, a flood – insurance coverage contributes to perpet-
uating society's vulnerability to climate change rather than to reducing it.[32]

These critiques have, however, done nothing to rein in the determination to expand the microinsurance business. Along the same
lines as microcredit in seeking to assure the 'financial inclusion' of the
poorest within capitalist finance, microinsurance is meant to allow
the reduction of the 'insurance gap', which prevents the greater part of
the population from paying its share into private insurers' profits. By
insurers' own estimation, today the world is massively 'under-insured',
and if developed countries represent close to 83 per cent of insurance
premiums paid worldwide (an amount close to $4.65 trillion, or 6.3
per cent of world GDP) the emerging countries count for only 17 per
cent, despite containing the majority of the global population.[33] Even
within rich countries, what the companies see as large 'pockets of
under-insurance' remain.

Thus, insurers have proposed a dozen measures in order to
'address the insurance protection gap' and combat 'under-insurance.'
The three first ones are:

1. Financial literacy programmes jointly funded by insurers, advisors and governments . . .

---

31 Matthieu Leimgruber's expression, in his *Solidarity Without the State? Business and the Shaping of the Swiss Welfare State, 1890–2000*, Cambridge: Cambridge University Press, 2008.

32 Paul O'Hare, Iain White and Angela Connelly, 'Insurance as Maladaptation: Resilience and the "Business as Usual" Paradox', Environment and Planning C, vol. 34, no. 6, 2016, pp. 1175–93.

33 Kai-Uwe Schanz and Shaun Wang (eds), *The Global Insurance Protection Gap: Assessment and Recommendations*, Geneva: Geneva Association, 2014, p. 5.

2. Microinsurance to cater to the needs of 4 billion uninsured people...

3. Public–private partnerships (PPPs) to address the protection gap.[34]

Looking at things through these lenses, the UNEP and insurers present global warming as the latest risk. And they vaunt the private insurance sector's expertise as the factor making it best prepared to guide adaptation.[35] Swiss Re, the second biggest reinsurance firm, has for some years promoted microinsurance as an adaptive solution to climate change. It enthusiastically predicts that microinsurance can be the main risk-covering strategy for the poorest billions of humans who are currently excluded. This determination to extend the scope of financial services is not a matter of charity, but expresses the need for this sector to build new markets in order to ensure its own profitability. As Swiss Re writes:

The microinsurance market could generate up to $40 billion in premiums. Over the last ten years, NGOs, insurers and community organisations have launched microinsurance programmes in several big markets and ranges of products. The growth of this business is mainly stimulated by the increased penetration of microfinance (especially microcredit), active government participation in some markets, and a supply of products based on needs.[36]

The estimate of a potential $40 billion offered in this document shows that the 'commercially viable' microinsurance market is aimed at population sectors with an income of between $1.25 and $4 a day; this market is estimated at 2.6 billion people and $33 billion in

---

34   Ibid, pp. 9–10.

35   UNEP Finance Initiative, *Insuring Climate Resilience: How Insurers Are Responding to Climate Change*, Geneva: UNEP, 2013.

36   Swiss Re, 'La Dernière étude sigma de Swiss Re souligne l'importance et le potentiel de la microassurance pour favoriser le développement socio-économique des marchés émergents', press release, 15 December 2010, media.swissre.com.

premiums. But the reinsurer goes further, as it envisages that the frontier of insurance can be extended yet lower down the income scale. The global population earning less than $1.25 a day encompasses another 1.4 billion people, who could generate up to $7 billion in premiums! Yes, these are very small amounts relative to the whole global insurance sector. But insurers take such a close interest in these markets because they are rapidly developing, and should – in the long term – ensure growth possibilities.

Nonetheless, the poorest of the poor will not by themselves manage to contribute to insurers' profits. For that, the state's role is indispensable. Only on condition that the state subsidises the destitute will they be able to subscribe to microinsurance premiums which allow them to 'cover their risks'. Thus, rather than imagine a state providing a universal, public insurance service (that is, social insurance), insurers want the state to be limited to its role of developing a capitalist rationality and subsidising their own activities. The UNEP describes this vision in insurance newspeak: 'There will always be some residual risk after a process of risk control, and, in the context of making risk transfer accessible to vulnerable communities, insurers believe the best insurance approach is a public–private collaboration.'[37]

The problem is, public–private partnerships (PPPs) do not work. Not, at least, from the viewpoint of the good management of public resources. When it comes to insurance, as with so many other sectors subjected to PPP ideology, state provision of universal public services remains infinitely superior.[38] This inconvenient truth has not stopped the UNEP any more than the other international organisations (like the World Bank) engaged in the full-tilt promotion of the private sector as the main solution to developmental and environmental problems. Subordinated to the development of private-sector commercial operations, the state's role in creating and sustaining the microinsurance market is, however, an indispensable one – as

---

37   UNEP FI, *Insuring Climate Resilience*, p. 6.
38   David Hall, *Why Public–Private Partnerships Don't Work: The Many Advantages of the Public Alternative*, Geneva: Public Services International, 2015.

insurers themselves recognise. They thus affirm the fact that in our neoliberal era, one must 'govern for the market, rather than because of the market'.[39]

However, beyond the question of the subsidies necessary for its creation, this microinsurance market cannot arise spontaneously from some unsatisfied demand among the poorest populations. The – solvent – demand for insurance must be actively prompted, produced and maintained. Similarly, the supply of financial products must be carefully calibrated for the particular conditions of the targeted populations. Insurers have thus mounted full-scale trials with microinsurance systems in order to develop and calibrate their products. And adaptation to climate change has itself served as their lever to carry off this 'penetration of microfinance', evoked by Swiss Re.

Backed by the Rockefeller Foundation, Swiss Re has thus developed a 'risk transfers' pilot programme directed at adaptation to climate change, labelled HARITA. It has done this in close collaboration with the United Nations World Food Programme (WFP), the charity Oxfam America and Columbia University's Earth Institute (headed by Jeffrey Sachs). Located in Ethiopia, this trial run offered microinsurance services to nearly 13,000 households between 2009 and 2011, allowing them to insure their harvests against climate hazards.[40] Encouraged by the success of this first initiative, Swiss Re and its partners (WFP, Oxfam, and the American government's USAID development programme) decided to make it permanent, under the title 'R4 Rural Resilience Initiative: Partnership for Resilient Livelihoods in a Changing Climate'.[41]

The imbrication of the different forms of microfinance and, more generally, the entrepreneurial philosophy of risk management which

39   Michel Foucault, *The Birth of Biopolitics: Lectures at the Collège de France, 1978–79*, Basingstoke, UK: Palgrave Macmillan, 2010, p. 121.

40   Swiss Re, 'Pilot Microinsurance Program Has a Successful Payout to over 1,800 Ethiopian Farmers After Drought', press release, 17 November 2011, swissre.com.

41   Swiss Re, 'R4 Rural Resilience Initiative: Partnership for Resilient Livelihoods in a Changing Climate', brochure at media.swissre.com.

informs it, clearly shines through from the project's objectives: 'The project (known as R4) is unique in developing a holistic risk management framework that includes risk reduction, risk transfer (insurance), prudent risk taking (credit), and risk reserves (savings).'[42]

When such objectives are spelled out, we can understand that the aim is not just the creation of an institution, but beyond that the creation of new behaviours based on a capitalist economic rationality. Here, poor peasants' way out of poverty, or at least insecurity, will depend on their capacity to adopt a strategic and entrepreneurial world view. It is this which is meant to lead them to see their day-to-day choices as decisions on the allocation of rare resources which will allow them, in managing their risks, to maximise their economic utility.[43] Hence poverty is no longer anything but the consequence of bad choices, a failure to save, or an overly risky (or else *insufficiently risky* and thus insufficiently productive) allocation of their meagre 'capital'.

EXPERIMENTS IN MICROINSURANCE

How does such a pilot programme actually work? The anthropologist Nicole Peterson studied its roll-out in a village in the Tigray region of northern Ethiopia between 2007 and 2010, where 95 per cent of the 4,000 inhabitants depend on agriculture.[44] Ethiopia, one of the poorest countries on the planet, frequently suffers droughts and famines. The bulk of its agriculture depends on rainfall to provide water for

---

42 Oxfam America, 'Innovation Case Study: R4 Rural Resilience Initiative', policy-practice.oxfamamerica.org.

43 This approach also runs into an impasse because of the fundamental problem that the power relations that allow the (epistemological) designation of some territory or event as 'risky' themselves constitute it as an (ontological) risk. See Julien Rebotier, 'Vulnerability Conditions and Risk Representations in Latin America: Framing the Territorializing Urban Risk', *Global Environmental Change*, vol. 22, 2012, pp. 391–8.

44 Nicole D. Peterson, 'Developing Climate Adaptation: The Intersection of Climate Research and Development Programmes in Index Insurance', *Development and Change*, vol. 43, no. 2, 2012, pp. 557–84. The paragraphs that follow are a summary of this article.

irrigation, and it is thus heavily vulnerable to climatic variations. Ethiopia is also a privileged site of intervention for international development aid programmes. It thus provides the 'ideal' conditions for the implantation of a climate microinsurance pilot programme. This kind of trial moreover follows in the longue dureé history of colonial and postcolonial interventions in Africa. As the geographer Michael Watts remarks, 'Africa is a laboratory for how to adapt to the potentially devastating impacts of climate change, or one might say life in the Anthropocene.'[45]

The peasants in the village studied by Peterson practice both pluvial agriculture (for instance, maize or teff, a local cereal) and agriculture dependent on irrigation (tomatoes, oranges, etc.). Far from being equally shared out, the land is distributed according to a sharp social hierarchy, in which some possess the largest, best, irrigated land holdings, while others (essentially meaning women) possess smaller strips without access to water. In this difficult social and ecological environment, survival strategies during drought periods are limited. Poor peasants can have recourse to food aid provided by the government in exchange for working on community projects. Often they are forced to sell their livestock – a strategy that increases their vulnerability in the following season, for livestock are useful for working the fields as well as for manure, providing an essential fertiliser. Loans are another possible solution, but they often mean long-term indebtedness. Peasants choose to diversify what they produce, for instance turning to plants which require less water or ripen later in the season. But this freedom of choice is itself limited by the size and quality of the land available to them. Thus, social differentiation operates at full tilt. The owners of irrigated land are much less vulnerable to rainfall variations than those whose land is entirely dependent on rain. As for landless peasants employed in sporadic farm jobs, they enjoy almost no margin of security.

---

45 Michael Watts, 'Adapting to the Anthropocene: Some Reflections on Development and Climate in the West African Sahel', *Geographical Research*, vol. 53, no. 3, 2015, p. 291.

In this context, the microinsurance pilot programme proposed to cover the losses to teff production in the case of extreme climate events. This programme was first of all addressed to the poorest, in essence women who farmed non-irrigated land. Since this was a trial run, the premiums were subsidised by international sponsors and not paid in cash by the participants. They did, however, have to provide something in return – namely, working for community projects (for example, planting trees or clearing out irrigation channels).

Other than requiring payment 'in kind' of these premiums, these microinsurance projects were characterised by their index-based mode of functioning. While classic insurance systems rest on the idea of compensating for some real damage (and thus require a monetary evaluation of the harm caused), insurance based on indices (or 'parametrics') frees itself of this constraint. The compensation is paid out to the beneficiary once the chosen index has reached some threshold determined in advance. Compensation is thus triggered automatically once certain temperatures are reached, or the level of rainfall in a given region drops below the threshold that has been decided. The insured parties do not have to provide any proof of the damage suffered; in extreme cases, a farmer thus insured could receive compensation even without actually having suffered any damage. Yet the opposite is of course also true: if a farmer loses their harvest without the index having been reached, they cannot claim compensation. Thanks to this system, insurers save on major costs in terms of evaluating damage. Most importantly, they immunise themselves against 'moral hazard' – in other words, the possibility that, because they know they are insured, the insured parties will adopt risky behaviour or even try to defraud the insurer by deliberately causing the damage that will trigger a payout.

The development of this type of insurance clearly makes it necessary to improve the accuracy of meteorological measurements, since these measurements are decisive in determining whether compensation is triggered. They not only have to be accurate and made available across the right geographical scale, but must also be credible – that is, transparent and comprehensible to the insured parties themselves.

But the places where insurers are seeking to develop microinsurance lack public infrastructure and, more importantly, weather stations. The involvement of the public authorities – demanded by the insurers themselves – is crucial in subsidising and maintaining a web of measuring devices, even though it is founded on the knowledge of experts from outside the country.

Similarly, the populations to whom microfinance is directed generally have a practical relationship with the climate rather than one mediated by abstract measurements (temperature, precipitation, wind strength and direction, humidity, etc.). The difficulty for insurers thus lies in both developing an 'education' in the insurance system (the famous 'financial literacy campaigns' mentioned above) and in creating confidence in the abstract measures on which compensation is based. In the case studied by Peterson, the involvement of development aid NGOs proved crucial in this stage of the process. These NGOs dispose of unrivalled know-how in terms of encouraging and structuring the community's acceptance – and negotiation – of the form in which meteorological information is represented and communicated.

Peterson shows the extent to which the microinsurance trial required the constant intervention of the insurers and their local linkmen (NGOs, etc.) as well as the subsidisation of the premiums by the project's international sponsors. She notes that the generalisation of a microinsurance system of this type carries great risks. As it is extended from 4,000 Ethiopian peasants to the planet's 4 billion 'under-insured' poor people, microinsurance can only lose the very characteristics which made it work at a local scale. Going global, in its bid to become a profitable activity, the system would have to leave behind the subsidies which allowed the poorest to take part in the pilot programme.

Building up climate security by entering an insurance market supposes (1) the adoption of a capitalist economic rationality; and (2) the development of a production output which can effectively be sold on a monetised agricultural market (in order to pay the insurance premiums, however 'micro-' they may be). The penetration of microinsurance thus effects an extension of market relations and the behaviours that they presuppose. However, in attaching this form of climate security to the success of the global insurance market, they create a further economic vulnerability for the populations dependent upon it: the vulnerability inherent in the market itself. Driven to commercialise and monetise their production (and thus to produce in a manner that anticipates market demand), peasants become more vulnerable to a collapse in prices than they would be in an economy essentially based on local consumption or else sustained by guaranteed prices for agricultural produce.

Above all, making climate security dependent on the financial markets is a big risk. The consequences that the US sub-prime crisis, with its trail of debt and evictions, had for the popular classes – especially for Blacks[46] – or the ones microcredit had for poor peasants in Andhra Pradesh, strikingly illustrate the possible consequences of the financialisation of everyday life –the very thing that the microinsurance model seeks to promote. Indeed, by concentrating on compensating peasants in the event of climate fluctuations, microinsurance ignores the fact that global warming not only brings greater uncertainty but also makes production conditions structurally more difficult.

The big global insurers' and reinsurers' intention is, without doubt, to increase their number of customer and risk portfolios, and, on this basis, to ensure their own long-term profits. But this does not explain the enthusiasm of the other actors involved in this project. What does a charity like Oxfam gain from this? How about an international organisation like the World Food Programme?

In the absence of a counter-movement to protect society, capitalism enjoys such ecological dominance that the institutions trying to find solutions to the problems of poverty or environmental degradation cannot do this except according to capitalism's own parameters. And that means constantly extending the logic of the market. Among

---

46   Gary Dymski, 'Racial Exclusion and the Political Economy of the Subprime Crisis', *Historical Materialism*, vol. 17, no. 2, pp. 149–79. See also David McNally, *Global Slump*; Soederberg, *Debtfare States and the Poverty Industry*.

the various solutions mapped out to deal with global warming, only the ones that reinforce the dominant neoliberal rationality have any chance of being selected and promoted by international sponsors. For Michael Watts, 'Building African resiliency in drought-prone regions is in profound ways a sort of Hayekian project: how a spontaneous market order will be built from and out of individual and community self-making and self-regulation through means of calculation and commodification.'[47]

Ignoring the social conditions that have produced and continue to produce the vulnerability of the poorest peasants (processes of exploitation, marginalisation, inequalities of power, etc.) the proponents of microinsurance seem to be offering a pragmatic, non-ideological solution to the increased insecurity produced by global warming. But this solution, relying as it does on the extension of the market, creates new social and economic vulnerabilities.

---

47   Watts, 'Adapting to the Anthropocene', p. 295.

Chapter 4

Climate Migrants: From Security Threat to Entrepreneurial Instrumentalisation

> In the time of fables, after the floods and deluges, armed men rose
> from the ground and exterminated each other.
>
> Montesquieu[1]

In a warlike speech, Barack Obama warned the Republicans who
continued to block any advance on the climate front that global
warming is a danger to US security. The president was clear about the
specific nature of this danger: it threatened an influx of 'climate
change refugees – and I guarantee the Coast Guard will have to
respond.'[2]

Far from marking a turning point in US security policy, the fear of
displaced populations resulting from the destruction of the environment is a recurrent theme. Already in the 1950s the Pentagon and the
US secret services saw environmental crises as a threat – and indeed, as
a weapon – in what they saw as the ineluctable conflict with the Soviet
Union. In 1969, Richard Nixon officially announced the US government's intention of treating environmental questions as a security issue.
He even suggested – to his allies' great consternation – that NATO
should concern itself with environmental matters.[3]

---

1 *The Spirit of the Laws*, XXIII.23.
2 Cited in Julie Hirschfeld Davis, 'Obama Recasts Climate Change as a Peril with
Far-Reaching Consequences', *New York Times*, 21 May 2015, p. 14.
3 Hamblin, *Arming Mother Nature*, p. 192–3. Razmig Keucheyan offers a detailed
discussion of the armed forces' approaches to climate warming in chapter 3 of his *Nature Is
a Battlefield*. On the relations between security and the environment at the international

In 1974, the CIA commissioned a report on the security implications of climate change. This rather strange document is saturated in the neo-Malthusian theses that were then in vogue (especially those of Reid Bryson and the Paddock brothers). Based on the theory of global cooling, the CIA report expressed worry over the droughts of the early 1970s and the looming possibility of famines which, it was supposed, would produce conflicts and political collapse. It concluded that it was necessary to develop 'methodologies ... to project and assess a nation's propensity to initiate militarily large-scale migrations of their people as has been the case for the last 4,000 years'.[4] It moreover regretted that the United States did not have a sufficiently developed climate research programme to forecast the future evolutions of the climate.

Fifteen years later, the young senator Al Gore could declare:

> Some may believe that the idea of the environment as a national security issue is just rhetoric. But we also know that just as the world has been living with the possibility of man-made disaster in the form of nuclear war, so it now lives with the growing threat of man-made disaster in the form of catastrophic environmental failure.[5]

Al Gore never did become president. And in 2004 the man who had stolen the election from him thanks to the manoeuvring of his Supreme Court friends, George W. Bush, preferred to disappear the

---

stage, see Lucile Maertens, 'Le Défi de la sécurité environnementale à l'ONU', in François Gemenne (ed.), L'Enjeu mondiale: L'Environnement, Paris: Presses de Sciences Po, 2015, pp. 205–14. For a piece putting the 'security threat' approach to climate migration into perspective, see Betsy Hartmann, 'Rethinking Climate Refugees and Climate Conflict: Rhetoric, Reality and the Politics of Policy Discourse', Journal of International Development, no. 22, 2010, pp. 233–46.

4   A Study of Climatological Research as it Pertains to Intelligence Problems, Washington, DC: Office of Research and Development, Central Intelligence Agency, August 1974, p. 31.

5   Albert Gore Jr., 'The Global Environment: A National Security Issue', in Ruth DeFries and Thomas S. Malone (eds), Global Change and Our Common Future: Papers from a Forum, Committee on Global Change, National Research Council, Washington DC: National Academy Press, 1989, p. 185.

Pentagon's apocalyptic report on the climate, which predicted millions dead and infinite disasters.[6] Despite this, the idea that global warming would lead to wars and waves of climate refugees remained embedded in people's minds, including at the highest levels of international politics. Sadly, this idea is a toxic one.

MIGRATION: A CONSEQUENCE OF CLIMATE CRISIS?

When specialists in the natural sciences began to try predicting the impacts of climate change, in the late 1970s and early 1980s, it seemed obvious to them that one consequence would be migration. If drought left land unsuitable for growing crops, the rural communities who depended on this land would simply have to move elsewhere. Rising sea levels, on account of the thermal expansion of the oceans and the melting of the polar ice cap, would automatically force those living near the coast to move to less precarious territory. The increase in climate variations, producing droughts, floods and other natural disasters, would also surely lead to massive movements of displaced populations.[7] Land ownership, the disruption of modes of life and of production, and the actual costs associated with migration, rarely even catch these authors' attention.

While this thinking is being *applied* to something new – global warming – it follows in a long scientific tradition analysing the relationship between environment and migration, dominated by a determinist vision and informed by a neo-Malthusian logic.[8]

---

6   Mark Townsend and Paul Harris, 'Now the Pentagon Tells Bush: Climate Change Will Destroy Us', *Guardian*, 22 February, 2004.

7   For instance, in the work of climatologists Reid A. Bryson and Thomas J. Murray, *Climate of Hunger*; or in Nels Winkless and Iben Browning, *Climate and the Affairs of Men*, New York: Harper's, 1975.

8   For a history of these debates, see Étienne Piguet, 'From "Primitive Migration" to "Climate Refugees": The Curious Fate of the Natural Environment in Migration Studies', *Annals of the Association of American Geographers*, vol. 103, no. 1, 2013, pp. 148–62. See also Romain Felli, 'Environment and Migration', in Noel Castree et al. (eds), *International Encyclopedia of Geography: People, the Earth, Environment, and Technology*, New York: Wiley-Blackwell and the Association of American Geographers, 2016.

For the climatologist William Kellogg – father of the WMO's first global climate programme – and his political scientist colleague Robert Schware, like him a veteran of the Annapolis seminar, it is self-evident that populations affected by climate change will try to move, in order 'to seek as nearly the same climate as possible to what one now has – a subjective feeling as to one's "comfort zone." '[9] They nonetheless warn against the potentially destabilising effects of these migration movements, which could entail political and social friction. But happily, not all these movements are problematic! Moreover, 'Canada coping with a possible American migration of several million people looks like one of the more easily resolved cases from a political standpoint alone'.[10] To my knowledge, even thirty-five years later, there is no concrete evidence that this is happening.

Writing in the same era, the philosopher Klaus Meyer-Abich remained rather more sober, relativising the impact that global warming would have on migration. The real problems were still questions of economic development (and overpopulation); global warming only added an extra layer to an already complex situation, and there was no reason to treat it differently from the problems that already existed.[11] The political scientist Dean Mann suggested that one adaptation strategy could consist of 'the offering of incentives to those whose behaviour may affect the climate or affect the behaviour of those who are affected by the climate'.[12]

But it was a document from UNEP, authored by Essam El-Hinnawi, which brought this question into the international political arena in 1985.[13] In this report, simply titled *Environmental Refugees*, the UN expert detailed three types of victims of environmental transformations: first, those who were temporarily displaced

---

9   William W. Kellogg and Robert Schware, 'Society, Science and Climate Change', *Foreign Affairs*, vol. 60, no. 2, 1982, p. 1098.

10   Ibid.

11   Klaus M. Meyer-Abich, 'Socioeconomic Impacts of CO_2-Induced Climatic Changes', p. 383.

12   Mann, 'Research on Political Institutions and Their Responses', p. 139.

13   Essam El-Hinnawi, *Environmental Refugees*.

on account of some extreme natural event (floods, earthquakes, etc.); then, those who had to migrate and make their homes in some new environment, on account of development projects; and lastly, the victims of a slow degradation that made their environment unlivable, as in the cases of desertification and deforestation (and potentially, global warming).

If 'environmental refugees' were thus painted as victims of global environmental transformations, they were themselves assigned part of the blame. For in this reading, the poor are exercising unsustainable pressure on their surroundings by 'over-exploiting' the available resources. So, for example, the poor would be responsible for deforestation, since wood (the 'biomass') is necessary for their heating or for cooking their meals. The resulting environmental degradation thus stems from the poor and their poverty.

El-Hinnawi popularised another neo-Malthusian theme that would enjoy a certain success over the next two decades: the idea that poor migrants destroy the environment which they come and settle in. Thus, multiple pieces of research were conducted in the 1980s and 1990s regarding the environmental impacts of refugee camps, their contribution to the degradation of ecosystems and their role in the possible outbreak of violent conflicts over resources.

This neo-Malthusian 'degradation narrative', as the sociologist Betsy Hartmann calls it, mobilises old colonial stereotypes and presents the populations of the Third World as too numerous and careless with regard to their environment. This, in turn, is used to justify Western interventions – in yesteryear carried out in the name of civilisation, and today in the name of saving the environment. Most importantly, this narrative 'blames poverty on population pressure, and not, for example, on lack of land reform or off-farm employment opportunities; it blames peasants for land degradation, obscuring the role of commercial agriculture and extractive industries and it targets migration both as an environmental and security threat'.[14]

---

14   Betsy Hartmann, 'Rethinking Climate Refugees and Climate Conflict', p. 234.

CLIMATE MIGRANTS: THE NEW BARBARIAN HORDES

In an intervention in the pages of the influential magazine *Foreign Affairs* just after the end of the Cold War, Jessica Mathews did not show similar coyness. For this organic intellectual of the US strategic apparatus, the degradation of the environment, for which the countries of the Global South were themselves responsible (especially through their demography) represented the new existential threat to the US superpower. And 'environmental refugees' were part of this threat, because they 'spread the disruption across national borders'. Mathews chose the example of the Haitian boat people (again Haiti) she said would descend upon the US coastline as they fled their now unlivable homeland.[15]

Where El-Hinnawi put on the guise of a certain humanitarian benevolence, Thatcher advisor Crispin Tickell, like Mathews, provided a more directly political analysis of the fear of environmental refugees. The inventor of the notions of 'eco-refugees' and 'environmental wars' stayed true to his neo-Malthusian logic.[16] Climate change would set in motion hordes of hungry poor people, who would come knocking at the door of the rich countries, dragging desolation along with them.

As we have seen, one of the problems of overpopulation, as Tickell presented it, was precisely that it would forestall adaptation to environmental changes:

In the past, when the world was a big place and the human population was a relatively sparse species, there was always a simple answer to climate change: people got up and walked somewhere else ... Now you have the same prospect, but in a very short space and time and in a heavily populated planet. There will be major difficulties because as you get a wave of refugees coming up from somewhere,

15 Jessica Tuchman Mathews, 'Redefining Security', *Foreign Affairs*, vol. 68, no. 2, 1989, p. 168.
16 Fred Pearce, *Climate and Man: From the Ice Ages to the Global Greenhouse*, London: Vision Books and LWT, 1989, p. 124.

there will be a man at the border asking for their passports. You can imagine very big problems beginning.[17]

Yet for Tickell, the problem of these supposedly massive population movements was above all a matter of the political economy underlying them. For it was *the poor* who would be descending on the Global North:

Is it going to be possible to resist this movement of refugees, of people looking for food and water and support and jobs and somewhere to put their families? Western Europe is a comfortable place where by and large things work pretty well. Yet if there were a big, big influx of refugees, would we be able to resist that influx on humanitarian grounds? And would you be able physically to resist them coming in? The USA cannot prevent the Mexican population from pouring across the frontier, even now.[18]

These refugees were a 'major threat' that could drive 'destabilisation' on account of their foreign mores and cultures, but also because they would supposedly bring disease. Worse, given the bankruptcy of the states of the Global South – incapable of controlling their populations – these refugees would, in Tickell's reading, be factors for disorder, terrorism and even economic collapse.[19] And what always shines through in Tickell is a fear of being swamped by the poor, which blends into a more properly racial fear:

People and governments could not, however, seal themselves off from the rest of the world. Land frontiers can always be penetrated. The movement northwards of Mexicans and other Latin Americans across the long southern frontier of the United States has so far

---

17   Tickell cited in Pearce, *Climate and Man*, pp. 124–5.
18   Tickell cited in Pearce, *Climate and Man*, p. 125.
19   Crispin Tickell, 'Risks of Conflict – Resource and Population Pressures', Linacre Lecture, University of Oxford, 8 March 2001.

proved irresistible, and every year parts of the United States develop more Hispanic characteristics.[20]

The spectre of the 'Hispanisation' of US territory is a recurrent theme among reactionary and conservative American elites. It has received scholarly validation in the 2004 book by the political scientist Samuel Huntington, the theorist of the 'clash of civilisations'.[21] For Huntington, Latin American immigrants are undermining the 'national identity' of the United States. This culturalist interpretative lens implies a fixed vision of national identities, anchored in specific territories and framed by sharp borders. The supposed problem of global warming, as neo-Malthusians see things, is that it sets in motion different ethnic groups, who are bound to enter into conflict with other groups over limited territory.

CLIMATE AND MIGRATION: THE US EMPIRE TOTTERS

Another neo-Malthusian who sees Mexican immigration as a threat to US security is Norman Myers, a conservation biologist who made his name with studies of biodiversity.[22] Between the early 1990s and the mid 2000s he published a number of reports, articles and books which left their mark on the debate on climate refugees. For instance, under the auspices of the Climate Institute, Myers published *Environmental Exodus*, a study financed, among others, by the Rockefeller Foundation and the US State Department – and with a foreword by Tickell.[23] The intention of this work was to draw the

---

20   Ibid. On the racism embedded within the discourse around climate refugees, see Andrew Baldwin, 'Racialisation and the Figure of the Climate-Change Migrant', *Environment and Planning A*, vol. 45, no. 6, 2013, pp. 1474–90; and Giovanni Bettini, 'Climate Barbarians at the Gate? A Critique of Apocalyptic Narratives on "Climate Refugees"', *Geoforum*, vol. 45, 2013, pp. 63–72.

21   Samuel P. Huntington, *Who Are We?: The Challenges to America's National Identity*, New York: Simon & Schuster, 2004.

22   Norman Myers, 'Environment and Security', *Foreign Policy*, no. 74, 1989, p. 32.

23   Norman Myers and Jennifer Kent, *Environmental Exodus: An Emergent Crisis in the Global Arena*, Washington, DC: Climate Institute, 1995.

attention of decision-makers and the general public to global warming and its consequences. Myers had long pleaded that the degradation of the environment should be understood as a problem that threatened US national security. He fuelled a cluster of security-related discussions that tried to present global warming as a problem of national security – one that could potentially lead to international wars – to try and push it to the top of the political agenda.

Polar bears dying off would not be enough to make that happen. Myers proposed instead to foreground the human victims of global warming, the 'climate refugees'. In 1995 he issued a shock figure: 50 million refugees by 2010, number that could eventually rise to 200 million.[24] Far from just the ruminations of one isolated scholar, Myers's numbers stimulated public debate and the international political agenda. The UNEP circulated his research, which was even included in the Stern report (so named after its author, a former chief economist at the World Bank) published by the UK government in 2006 in an attempt to measure the economic costs of global warming.

This number of 200 million refugees was bound to stoke fear. But by comparison, the total number of political refugees worldwide in 2013, according to the High Commissioner for Refugees (UNHCR), was barely 18 million. While this was not necessarily Myers's intention, under the cloak of humanitarianism this vision played on xenophobic sentiments by more or less openly mobilising images of massive influxes, barbarian invasions, hordes of the destitute descending on the countries of the Global North in quest of more hospitable climes.[25]

In the political context of the 2000s, the logical consequence of this rhetoric was not to open the borders and welcome in 'all the world's misery'[26] which resulted from climate disruption. Rather, it

---

24   For instance, Myers and Kent, *Environmental Exodus*, pp. 23–41. See Hartmann, 'Rethinking Climate Refugees and Climate Conflict', pp. 235–6.

25   See Bettini, 'Climate Barbarians at the Gate?'

26   A reference to French prime minister Michel Rocard's repeated assertion that 'France cannot take in all the world's misery', a line later repeated by Emmanuel Macron.

provided a further justification to strengthen the apparatus for the long-distance control of migrants, at the borders of the European Union and the countries of North America. In the European case, it has been strengthened with the active – even proactive – participation of certain countries on the southern shore of the Mediterranean, which have tried to position themselves politically as 'managers' of migration from sub-Saharan Africa to the EU – a role that can only grow as global warming continues.[27] The strategy for the 'securitisation' of climate-induced migration has thus succeeded in placing the subject on the political agenda. But it appears in a form which very much risks turning out hostile to the rights and interests of migrants by painting them as a threat to border security.

Above all, this strategy resonates with the rise of a post–Cold War vocabulary which seeks to cast the environmental catastrophe into the existential threat bearing down on all humanity, taking the place of the threat from the communist states. In the 1990s, the biologist Myers, the strategist Mathews and the diplomat Tickell were far from alone in connecting poverty, environmental degradation, migration and violent conflicts. For instance, Al Gore declared that 'the new enemy' – environmental degradation – 'is at least as real as the old' – the Soviets.[28]

In 1994, the journalist Robert Kaplan published an article in the US monthly *The Atlantic* which was due to have a lasting influence, predicting that, after the fall of the Soviet empire, geopolitical tensions around resources and the degradation of the environment would be the destabilising factors in global society, driving the collapse of the states of the Global South.[29] This state failure – with their inability to 'manage' their own populations – would lead to the proliferation of violence, organised crime, trafficking, etc., and in a second movement

27   Gregory White, *Climate Change and Migration: Security and Borders in a Warming World*, Oxford: Oxford University Press, 2012.

28   Gore, 'The Global Environment: A National Security Issue', p. 180.

29   Robert Kaplan, 'The Coming Anarchy: How Scarcity, Crime, Overpopulation, Tribalism, and Disease are Rapidly Destroying the Social Fabric of our Planet', *The Atlantic*, February 1994.

rebound on the 'way of life' of the states of the Global North, starting with the United States. Here we are in a conceptual universe very close to that of neoconservatives like Samuel Huntington and his 'clash of civilisations' (which, indeed, Kaplan cites), who read the world as structured by oppositions between more or less incompatible cultures.[30]

A FUNCTIONAL NEO-MALTHUSIANISM

For these environmental neo-Malthusians, the human population, and especially its growth in the countries of the Global South, explains the rising pressure on resources and thus the fact of most of the human species being kept in poverty. In this model, climate change reinforces pre-existing pressures. It is a 'threat multiplier'. For instance, by increasing the periods of drought, it may lead to 'water wars'. It may also trigger migratory movements or lead to violent territorial conflicts to secure resources, and indeed to 'state failure' in the Global South. Many people rightly seeking to draw attention to the consequences of global warming grasp at this explanatory model. Indeed, this is also a model amplified by media who find material therein to indulge their own penchant for reductionist science – for instance, in the popular works by the biologist and geographer Jared Diamond.[31]

But this current of thinking attributes the influence of climate, and climate change, on migration only when it is has to do with poor

---

30 Samuel P. Huntington, 'The Clash of Civilizations?', *Foreign Affairs*, vol. 72, no. 3, 1993, pp. 22–49. The expression 'clash of civilisations' had been formulated by a very conservative historian of Islam, Bernard Lewis, in another article for this same monthly: 'The Roots of Muslim Rage', *The Atlantic*, September 1990.

31 Of course, few researchers today make the case for monocausality and a narrow climate determinism. But the 'multicausal' framework they employ nonetheless seeks to isolate one particular variable (the environment or the climate), instead of seeking to understand the historical and political process which leads to the existence of particular socio-economic formations and the vulnerabilities which they produce. For a response to Diamond's excesses by anthropologists and historians, see Patricia A. McAnany and Norman Yoffee (eds), *Questioning Collapse: Human Resilience, Ecological Vulnerability, and the Aftermath of Empire*, Cambridge: Cambridge University Press, 2009. In particular, this work returns analysis of ecological disasters to the *longue durée* history of imperialism and colonialism rather than reduce them to the 'bad choices' human societies supposedly made.

people in the Global South. Very rarely do we hear it said that rich Americans who move to Florida for their retirement are 'climate migrants'. Similarly, the quest for a small house with a garden, which leads many city centre residents to migrate to suburbs to buy a villa, does not make them into 'environmental migrants'. These categories are almost exclusively mobilised to describe the supposed behaviour of dispossessed populations in the countries of the Global South, betraying how far (post)colonial imaginaries are still well-embedded in the consciousness. Delimiting a sort of rising scale of civilisation, social and economic explanations are reserved to the North, while in the South the decisive thing is nature (even if it is modified by humans). Here we are not far from Montesquieu, when he wrote that 'Nature and climate almost alone dominate savages; the ethos used to set the tone in Lacedæmon; government maxims and the ancient ethos set it in Rome'.[32]

Forgotten, here, is everything to do with colonial and postcolonial history: the violent dispossession of people from their land, the submission of populations to an extractive labour regime, neoliberal structural-adjustment policies, the behaviour of multinational corporations, local power struggles. If it is global warming that produces poverty, displaced populations or even wars, then human institutions, political conflicts and socioeconomic inequalities will decline in relative importance. Nature – even if it is a nature which humans have involuntarily modified – becomes the great force determining the fate of human societies, a force standing beyond the possibility of political action.[33]

Even more than the fall of the USSR – though it is also linked to this – the political context which overdetermines these apocalyptic visions is that of the questioning of Southern countries' political sovereignty by imperialist countries. Emancipation from colonial rule and the independence of Third World countries (and their attempt to

---

32   *The Spirit of the Laws*, XIX.4.
33   Neil Smith, *Uneven Development: Nature, Capital, and the Production of Space*, Athens: University of Georgia Press, 2008, afterword to the 3rd edition, p. 245.

take back control over their own resources) were a victory over the interests of the colonial powers and their ruling classes. The narratives of environmental degradation are, in part, linked to a world view according to which the peoples of the Global South are incapable of self-government. Hence the rhetoric on weak states, failed states, human rights violations (real as these are), corruption, nepotism, the misappropriation of funds, etc.

In the context of the 1990s, the solutions which experts brought to these problems all revolved around good governance, the formation of reliable institutions and the extension of the market's domain. These institutions were meant to bring about a positive form of social control and economic development beneficial to all. In establishing clear property rights over resources, in guaranteeing the security of law – for international investors, that is – and reducing political instability, the countries of the South could hope to see a new era of growth that would respect the environment. Yet for many governments in the North, as for certain international organisations, these favourable conditions for the pursuit of economic accumulation would not materialise all by themselves. Constant outside interventions would be necessary if functioning states – lifted clear of local political struggles – were to be built.[34]

AN INGENUITY GAP?

Kaplan had drawn his apocalyptic description of failed states and populations on the move from a young political science researcher, Thomas Homer-Dixon. In the late 1980s, Homer-Dixon became head of a major University of Toronto research programme on 'environmental changes and violent conflicts'. According to Homer-Dixon, population pressures (that is, the fact that poor people in the countries of the South had too many children) destroyed natural resources, especially cultivable land and forests, and trapped already poor

34   David Chandler, *International Statebuilding: The Rise of Post-Liberal Governance*, New York: Routledge, 2010.

countries in poverty. These countries, which had a lesser capacity to adapt to environmental changes, and especially climate changes, squandered the few resources that they could have devoted to developing these capacities. And they thus weakened their populations' 'human security'.

The core of Homer-Dixon's theory can be reduced to the idea of 'ingenuity', a supposed characteristic of rich countries, but which is lacking in poor ones. Countries' capacity to adapt to climate shocks is thus measured by the yardstick of these differing levels of ingenuity:

> As more environmental problems and scarcities appear, and their interactions and synergies become more complex and unpredictable, policy-makers in developing countries will need greater and greater social ingenuity to design, implement, and maintain the social mechanisms (such as effectively operating markets) that could unleash the technical ingenuity required to address these problems. Simply put, these policy-makers will have to be increasingly clever as 'social engineers' just to sustain the current output of agricultural and other necessary goods.[35]

The greater ingenuity which the countries of the Global South could potentially bring to bear would allow them to adopt more efficient market mechanisms. We again find here the theme of uncertainty and unpredictability which we saw as basic to approaches that focused on 'resilience'. Indeed, those approaches also worried over the loss of populations' capacity to adapt and innovate,[36] preaching the gospel of flexibility.

Homer-Dixon pursued his endeavour of spiriting away the social causes of poverty in *The Ingenuity Gap*,[37] published in 2000. Notably, he boasted of having been consulted by Vice President Al Gore, in order to study solutions to the environmental causes of the main

---

35   Thomas Homer-Dixon, 'Environmental Change and Human Security', *Behind the Headlines*, 1991, vol. 48, no. 3, p. 10.

36   Holling, 'An Ecologist's View of Malthusian Conflict'.

37   Thomas Homer-Dixon, *The Ingenuity Gap*, New York: Knopf, 2000.

contemporary conflicts. An explanation that blamed an absence of ingenuity made it possible to ignore the social and historical causes of a lack of capacity to adapt to climate change, to such a point that a commentator like political scientist Eric Ross could write that it reminded him of 'the colonialist's disdain for the limited capacity of the colonised'.[38]

Thomas Homer-Dixon threw himself into a series of fantastical predictions based on the supposed aggressiveness of Southern countries (especially revolutionary regimes) which, prey to a lack of resources, would inevitably launch wars of territorial aggression. It is rather arresting that this kind of prose could be written at the very moment that the United States (which, was of course not deemed to be lacking in ingenuity) launched its own war of military aggression against Iraq, in order to control the oilfields there . . .

Homer-Dixon's works were the missing link between the narrowly security-focused, neo-Malthusian vision of environmental migration and a benevolent form that focused on technical solutions to ecological problems. But what he called ingenuity (which also implied technological and financial transfers from rich countries to poorer ones) fundamentally rested on the unleashing of the creative powers of the market and the gospel of social, institutional and economic flexibility.

THE FIRST CLIMATE WAR?

In early 2007, the UN secretary general Ban Ki-moon churned out an editorial for the *Washington Post* in which he explained that the conflict in Darfur was the result of global warming. In reducing rainfall in the region, global warming had resulted in a collapse in resources and aggravated territorial competition between nomadic herdsmen and sedentary crop-growers.[39] He thus adopted the thesis

---

38 Eric B. Ross, *The Malthus Factor: Population, Poverty and Politics in Capitalist Development*, London: Zed Books, 1998, p. 205.

39 Ban Ki-moon, 'A Climate Culprit in Darfur', *Washington Post*, 16 June 2007.

of the journalist Stephan Faris, as expounded in an article for the *Atlantic* a few months previously, and which did the rounds in international institutions like the UNEP. It was also used by the usual suspects of humanitarian neoliberalism like Jeffrey Sachs,[40] whose explanations blithely mixed the country's persistent 'poverty', the degradation of the local environment and global warming.

Yet, as the political scientist Harry Verhoeven notes, 'the violence in Darfur is anything but a Malthusian collapse: it is a deadly cocktail that had been building for decades through the fusion of power politics in Khartoum, political-economic marginalization of Darfur and local frustrations due to artificially created scarcity.'[41]

While not denying the existence of climatic fluctuations, Verhoeven shows that local populations' *vulnerability* to them is the result of a long history of political and economic processes which have shaped the means and strategies available to peasants to deal with these changes. Narratives of famine and conflict, and indeed of abundance (Darfur has in different eras been painted as the breadbasket of east Africa or even of the Middle East) have existed since the days of British colonisation – but mask this region's profoundly unequal development since that period.

In concentrating financial investment and hydraulic equipment in the Khartoum region and the Nile Valley, local elites, backed by the colonial administration, marginalised peripheral regions of Sudan, including Darfur. Even after independence, the prevalent political vision continued to respond to the needs of the global market by promoting agriculture oriented to cotton exports (very liable to price fluctuations on the global market) to the detriment of agriculture producing food for consumption. Colonel Nimeiry's violent seizure of power in 1969 was accompanied by a more 'developmentalist'

---

40   Japhy Wilson, *Jeffrey Sachs: The Strange Case of Dr Shock and Mr Aid*, London and New York: Verso, 2014.

41   Harry Verhoeven, 'Climate Change, Conflict and Development in Sudan: Global Neo-Malthusian Narratives and Local Power Struggles', *Development and Change*, vol. 42, no. 3, 2011, p. 692. On Darfur, the following paragraphs take their lead from Verhoeven's analysis.

vision grounded in big public investments and the building of mass-scale water-supply infrastructure. This was supposed to allow the 'modernisation' of Sudan and its agriculture. But this sudden valorisation of capital-intensive mechanised agriculture helped to marginalise peripheral communities and the poorest. By consolidating and mechanising the big agricultural properties, elites dispossessed the small peasants of common pasturelands and their transhumance corridors. Megaprojects diverting water from the Nile to irrigation, and the arrival of capital from the Gulf states, accentuated the reorientation of Sudanese agriculture towards producing exports for international markets.

This agriculture modernisation project would ignore all ecological recommendations – and it proved to be an environmental disaster, especially through exhaustion of aquifers and cultivable soil. Ignoring the example of the agrarian reforms in Asia (which involved land redistribution, an emphasis on agriculture aimed at local consumption and equipping small peasants with modern technologies), Sudan's modernisation programme precipitated famine conditions in the mid 1980s, as drought hit the country. During this crisis – here following a pristine market logic – grain continued to be exported from Sudan to be sold on the world market, even as people were dying of hunger in Darfur and humanitarian food aid was at the same time being sent there! The development model imposed by the country's elites greatly increased marginal populations' vulnerability to climate change.

But in identifying global warming as the trigger for the conflict in Darfur, Ban Ki-moon perhaps hoped – however clumsily – to give concrete expression to the threat it posed. But in so doing, he passed over in silence the much more concrete conditions which ensured that climatic variations would have political consequences. As Verhoeven notes, the sources of conflict are to be found not in environmental changes, but in the long history of power struggles in this country, of economic appropriation by the local elite and of the economic strategies that the country has followed.

A DEPOLITICISING HUMANITARIANISM

Fortunately, the question of climate refugees is not posed only in terms of security. Many voices in national parliaments and NGOs – and some public intellectuals – have spoken up in defence of climate refugees, holding that they are indeed victims. In particular, numerous jurists have proposed adoption of a new international convention – or new international norms – aimed to protect the victims of environmental and climate degradation.[42] Since the early 2000s, the Green group in the European Parliament has advocated recognition of these refugees' situation and proposed protections for them.[43] Climate refugees' fate has also inspired artists, photographers and filmmakers seeking to give visibility to the consequences of global warming. However, attempts to establish recognition of the existence of environmental or climate refugees continue to run into multiple legal and institutional obstacles.[44] For instance, the UN High Commissioner for Refugees and the International Organization for Migration (IOM) are competing to lay claim to this issue – and to nab the funding reserved for this question now in the public spotlight.

Paradoxically, this upsurge of solidarity depoliticises the climate question. With the build-up of catastrophist discourse on the

---

42   I have found the most interesting contributions to be Laura Westra, *Environmental Justice and the Rights of Ecological Refugees*, London: Earthscan, 2009; and Jane McAdam, *Climate Change, Forced Migration, and International Law*, Oxford: Oxford University Press, 2012; and in French, Christel Cournil and Benoît Mayer, *Migrations environnementales, enjeux et gouvernance*, Paris: Presses de Sciences Po, 2014.

43   Aurélie Sgro, 'Towards Recognition of Environmental Refugees by the European Union', *Asylon(s)*, no. 6, November 2008, Exodes écologiques, online at reseau-terra.eu.

44   There is a vast literature on this subject. For a detailed overview, see Romain Felli, 'Environment and Migration'. For good discussions of this process, see François Gemenne, 'How They Became the Human Face of Climate Change: Research and Policy Interactions in the Birth of the "Environmental Migration" Concept', in Etienne Piguet, Antoine Pécoud and Paul de Guchteneire (eds), *Migration and Climate Change*, Cambridge: Cambridge University Press, 2012, pp. 225–59; and James Morrissey, 'Rethinking the "Debate on Environmental Refugees": From "Maximalists and Minimalists" to "Proponents and Critics"', *Journal of Political Ecology*, 2012, no. 19, pp. 36–49.

consequences of global warming and the duty to aid climate refugees/ migrants, added weight is given to the idea that every problem stems from this damned global warming. This is not deliberate, but it has very real consequences. The immediate causes – wars, violence, dispossession, the processes of exclusion and marginalisation – disappear behind the grand narrative of global warming, to which is attributed an omnipotent role in determining human affairs. Refugees are hence seen as simple passive victims of a process (warming) which they could never cope with. Vulnerability to global warming is attributed to the whole of a territory or a population, without understanding that it is precisely (internal and external) power relations that determine different individuals' and social groups' differing vulnerabilities, even within the same territory, especially along the axes of class, gender and race.

This depoliticisation reached its climax in 2013 when the Swiss and Norwegian governments launched a so-called 'Nansen' initiative aimed at promoting the international protection of persons displaced by natural disasters. This initiative emphasised the degree to which global warming will contribute to population displacements and how important it is to protect those displaced outside of their own national borders.[45] In this regard, it is encouraging that Norway is so concerned about climate refugees, given that it is one of the world's oil exporters and thus one of the major sources of the current climate crisis. As for Switzerland, its government constantly refuses to regulate the activities of the multinationals operating from its soil, that extract and trade in raw materials (one of the main sources of dispossession and displacement in many countries in the Global South).[46] For three decades, Switzerland has also repeatedly toughened the conditions for granting political asylum. It thus has unquestionable credentials when it comes to managing the protection of environmental refugees . . .

---

45   See nanseninitiative.org.
46   See the well-researched campaign by Déclaration de Berne, a Swiss NGO: rohma.ch.

But the analyst who insists on the more immediate causes that render populations vulnerable can help to repoliticise the situation. Marginalised populations can fight to win back their rights or, indeed, put existing ones to practical use, creating conditions that will allow them to live with global warming. Yet these demands suppose a change in local and international power relations – and, in this sense, make up part of the wider political struggles for social emancipation.[47]

ALL ARE AFFECTED, NOT ALL MOVE

Migration specialists Roger Zetter and James Morrissey offer this conclusion from the case studies they have conducted in Bangladesh and Kenya, two countries particularly afflicted by global warming: climate changes affect everyone, but not everyone migrates in response.[48] Their studies show how a single phenomenon (environmental degradation owing to global warming) produces very different results, depending on the local power configurations, resources and institutions. In Bangladesh, in the absence of a strong central state, power is heavily concentrated at the local level, in the hands of landowners. These latter profit from the high levels of state corruption and the state's lack of effectiveness, buying up the land of illiterate and resourceless peasants at low prices, or even directly appropriating it. The vast majority of the rural population is thus dispossessed of land and forced to cultivate less fertile ground – or sell their labour-power to the landowners.[49] Environmental changes like the erosion or salination of the soil also affect populations in a highly differentiated way. The dispossessed experience a collapse in output but cannot turn to using better land, while those who hold power locally can adapt by

47   Andrew Baldwin, 'Pluralising Climate Change and Migration: An Argument in Favour of Open Futures', *Geography Compass*, vol. 8, no. 8, 2014, pp. 516–28.

48   Roger Zetter and James Morrissey, 'The Environment-Mobility Nexus: Reconceptualising the Links Between Environmental Stress, Mobility and Power', in Elena Fiddian-Qasmiyeh et al. (eds), *The Oxford Handbook of Refugee and Forced Migration Studies*, Oxford: Oxford University Press, 2014, pp. 342–54.

49   Ibid., p. 345.

choosing better locations and changing their production (for instance, shifting from rice to shrimp – a choice that also reduces rural employment, since shrimp production requires less labour than rice production). The most vulnerable families will tend to send at least one of their number – especially young women – to the cities to try to diversify their sources of income, even if the conditions are no better there than in the countryside. The landowners' monopolisation of the land thus brings about the marginalisation of landless populations and their greater vulnerability to environmental changes. The property-owning classes' ability to buttress their own security in the face of environmental changes thus proceeds by way of their capacity to increase the insecurity of the dispossessed.

In Kenya, conversely, strong webs of community and cultural norms tend to mean that adaptations to environmental change will be the result of collective decisions. Beyond the extended family structure, complex socio-political institutions like councils of elders organise the distribution of land, access to water and access to grazing land for animals. Though these collective forms of managing resources are being eroded (by the penetration of external institutions like the central state, but also international NGOs), they are still important there and make it possible to spread out the impacts of fluctuations owing to climate variations, organising a more solidaristic form of adaptation. By way of example, Zetter and Morrissey describe the mutual aid that rural families offer each other during tough times, informal insurance systems and non-commercial loans that extend across multiple generations, and the mutualisation of resources in periods of dearth.[50] This social structure tends to generate forms of local adaptation to environmental changes, resulting in less migration than in the Bangladeshi case.

The processes of dispossession and heightening vulnerability in turn provoke resistance and reactions. In Bangladesh, for example, some impoverished peasants rally together and, backed by local NGOs, assert their right to occupy the land, against the claims of the

---

50 Ibid., p. 348.

big landowners. In Kenya, the distrust toward the central state, as well as its own forced rehousing practices, prompt some to defend the traditional institutions for managing resources. They valorise these latter anew, as a strategy for adapting to climate change – thus averting any need to migrate.[51]

In short, one cannot paint climate migrants as mere victims of a process taking place above their heads – even if one's intention is to protect them or to offer them asylum. In reality, the decision to migrate in the face of environmental transformation depends on a complex articulation of power, access to resources, norms, and institutions; taken together, these factors mould populations' differentiated vulnerability to climate changes. Reducing this vulnerability requires collective processes of organisation and struggle, to democratise the power structure and better share out resources.

THE TRIUMPH OF THE NEOLIBERAL SPIRIT: THE AID MARKET

After their role in security fears and in depoliticising humanitarianism, climate migrants are now experiencing a third life. This time they appear in the guise of entrepreneurs, deploying a skilful economic strategy in adaptation to climate change.

'Migration will not and cannot and should not be stopped. It's good for countries, and it's good for the millions of people who vote with their feet however many restrictions are put in their way. Who will be brave enough to lead the last liberalisation of capitalism and give to people the same freedoms as cars, as clothes, and as computers?',[52] concludes the managing director of the ODI, an influential British think-tank specialising in development. It is as if cars or computers were endowed with consciousness, telling them they have to make their way onto the market in order to sell themselves at the best price. This comparison of migrants' freedoms to the free

---

51   Ibid., p. 351.

52   Claire Melamed, 'Migration Is Capitalism's Unfinished Business – It Cannot and Should Not Be Stopped', Comment ODI, Overseas Development Institute, 12 May 2015.

circulation of commodities betrays the thinking of the ODI and the many other back rooms of neoliberalism. For them, migrants' freedom has no value except insofar as they are a factor of production. 'Laissez faire, laissez passer' remains the misleading watchword of the promoters of the global market in goods and persons. Global elites trying to promote the freedom to move in order to sell oneself often criticise national 'egoisms': that is, the short-term vision of politicians purportedly implementing protectionist 'close the borders' measures. Yet what also stands out here is their hatred for democracy. For, if politicians are ardent about closing the borders, or at least putting on the spectacle of doing so, this is said to be under pressure from their electorates, who are invariably ignorant of neoclassical economic theory. The populace of the Northern countries would, then, bear the greatest responsibility for xenophobic – what are called 'populist' – policies.

Listening to these ideologues, one could easily believe that, on the one hand, there are the good capitalist entrepreneurs who want freedom of migration and, on the other hand, egoistic national politicians, flanked by their electorates, who are doing everything they can to protect themselves against competition on the labour market. In this reading, migrants' insecurity would be the simple side-effect of attempts by 'locals' to guarantee their own economic security. This fable would be just fine – if it did not forget the fact that the 'good' entrepreneurs rather rarely take up the struggle for the regularisation of sans-papiers.

The Northern countries' migration policy is thus constantly caught between the two apparently opposed – but in reality complementary – poles of securitarian nationalism (close the borders) and economic neoliberalism (the free flow of labour).[53] Far from being opponents of neoliberalism, the xenophobic parties turn out to be its greatest allies. This apparent contradiction is resolved once we consider that the insecurity this xenophobic tendency creates for

---

53   Gregory Feldman, *The Migration Apparatus. Security, Labor, and Policymaking in the European Union*, Stanford: Stanford University Press, 2012.

migrants (rendering them 'illegal', liable to expulsion, invisible in public space, etc.) supplies a regime of workforce control favourable to neoliberalism (workers do in fact move around, but largely stripped of rights). The insecurity of labour creates security for capital. Insecurity for workers is in the best interest of most capitalists – and they understand as much. As a result, states themselves set out organise such insecurity.[54] The spectacle of border 'control', the defence of territorial sovereignty (identified with the sovereignty a suburban homeowner has over their own backyard), the highly publicised expulsions of persons judged to be 'illegal', the intense police control of non-whites – going so far as the building of 'walls', 'barriers', surveillance networks, etc. – all come in addition to the wretched piling-up of the lifeless bodies of migrants drowned in the Mediterranean and elsewhere.

Yet, in rejecting the securitarian pole, some defenders of migrants and experts in 'climate migration' have chosen to embrace the other pole of this contradiction. With the collapse of the sense of possibility that the communist world represented (a false hope, doubtless, but one that had real effects), and the parallel decline of social democracy, neoconservative and neoliberalism together delimit either end of a now-shrunken political terrain. Within this space, political choices are made with the objective of allowing more efficient capital accumulation. Debates revolve around the best means of organising this accumulation – and whoever wants to intervene effectively in these debates must dress up their arguments accordingly.

Just as some defenders of the environment have appropriated the language of the market to talk about ecology, some defenders of migrants' rights foreground the – real or supposed – economic benefits of migration. Freedom of movement is thus defended less as a fundamental human right and more insistently as a spur to growth. In

54   Susanne Ferguson and David McNally, 'Precarious Migrants: Gender, Race and the Social Reproduction of a Global Working Class', *Socialist Register*, no. 51, 2015, pp. 1–23; Kristin Surak, 'Guestworkers: A Taxonomy', *New Left Review*, no. 84, 2013, pp. 84–102; Immanuel Ness, *Guest Workers and Resistance to U.S. Corporate Despotism*, Urbana: University of Illinois Press, 2011.

this portrayal, international migration represents a 'triple win' – for the arrival country, whose workforce will grow; for the departure country, which migrants will regularly send remittances back to; and for the migrants themselves, who will see their living standards improve. The idyllic narrative of the 'triple win' allows the (symbolic) mitigation of the chronic shortage of public funds to aid development and the fiscal crises of the countries of the Global South. The World Bank and other international institutions have, moreover, realised as much; they present migrants' own remittances as the main – and most stable – form of financing development back home!

A PRIVATISATION OF SOCIAL REPRODUCTION?

Yet this newfound enthusiasm for promoting migration is a merely instrumental use of the climate crisis. It pretends, on this basis, to have resolved the crisis of social reproduction – but without actually doing so.[55] Feminist studies have highlighted that, outside the formal economic sector, there need to exist institutions and practices which allow 'social reproduction'. Before they become workers able to sell themselves on the labour market, human beings have to be socialised – brought up, educated, trained, fed, housed, washed, entertained, etc. Similarly, old age, accidents, disability, infirmity and illness demand (sometimes individual) care – in short, a 'reproductive' labour provided by others. These other people, whose labour is the necessary basis of all productive activity (that is, an economic activity that increases the value of capital) are generally women, and their work generally goes unpaid.[56] To adopt the neat formula used by the feminist economist Nancy Folbre, the invisible hand of the market requires the invisible heart of reproductive labour.[57]

55   Rahel Kunz, *The Political Economy of Global Remittances: Gender, Governmentality and Neoliberalism*, London: Routledge, 2011.

56   Diane Elson, 'The Economic, the Political and the Domestic: Businesses, States and Households in the Organisation of Production', *New Political Economy*, vol. 3, no. 2, 1998, pp. 189–208.

57   Nancy Folbre, *The Invisible Heart: Economics and Family Values*, New York: New Press, 2001.

Yet, if the effort – and cost! – of social reproduction essentially weighs on women's shoulders, part of it is socialised, or taken into state hands. Primary, secondary, technical and sometimes university education is now part of the state's remit in most countries, albeit with vast differences in terms of its organisation, financing and coverage. The provision of a literate workforce is clearly in capital's interest, whereas it represents a major cost for the countries who provide it, whether it is funded by fees or taxation. The same can be said of healthcare, social security and pensions.

The global migration regime developed since the end of the Cold War constitutes a 'global market in human labor-power', as Susanne Ferguson and David McNally explain.[58] It sends massive numbers of workers from the countries of the Global South to Northern ones (especially Europe, the US and Canada) but transfers them also among the Southern countries (notably to the Gulf States). These workers – most of them women – more often than not migrate in conditions of severe precarity and insecurity, conditions created by state institutions themselves. Whether they are 'illegals' or temporary, circular or seasonal migrants, they are in a situation of structural inequality, stripped of most of their rights and constantly subject to possible expulsion.

At the same time, they contribute to economic accumulation – and to social reproduction. Nurses, cleaning ladies, nannies, live-in carers for the ill and homeworkers provide the care work necessary for social reproduction in the Northern countries. Yet while they provide this labour, the costs of their own reproduction (their schooling, education for their children, care for their parents, etc.) are not taken on by the institutions of the countries where they live. They contribute to economic accumulation and social reproduction in the Northern countries but draw little or no benefit from public services or socialised spending (education, healthcare, pensions) acquired by these countries' other residents. In this way, capital in the importing countries saves considerable amounts of money on these migrant

---

58   Ferguson and McNally, 'Precarious Migrants', p. 3.

women's backs (in unpaid social wages), transforming them into superexploited workers and boosting its own profits.

Moreover, out of their meagre wages migrants send remittances to their countries of origin. These transfers serve to finance the social reproduction of family members who have stayed behind (their parents, their extended family, and, increasingly often, their children). This is especially common if they have come from countries which provide only the slightest social protections and where multinationals' activities almost totally escape taxation. The constitution of the global market in human labour power simultaneously entails the import of migrant workers, in conditions of maximum flexibility, and the (re)privatisation of the conditions of social reproduction. This is the charged context in which the utilitarian arguments in favour of the economic benefits of climate migration are rolled out.

MIGRATION AS A STRATEGY FOR ADAPTING TO CLIMATE CHANGE

This utilitarian realignment of the defence of migrants would, towards the end of the 2000s, be directly imported into the debate on climate refugees by various international organisations like the World Bank, the IOM and groups of experts and scholars. They had a simple idea: stop considering migration in negative terms, as a repercussion of environmental degradation, and instead see it as a form of adaptation. In short, it was necessary to stop talking about climate 'refugees' and instead encourage environmental 'migrants' to create their own solution to climate disaster.[59]

So, what if climate change-induced migration were not seen as a problem, but as in fact a solution? Wouldn't it then be necessary to encourage those entrepreneurial migrants who make the investment in moving in order to boost their capital – thus not only showing a fine capacity to adapt, but also helping create resilience? As one World

---

59 Romain Felli, 'Managing Climate Insecurity by Ensuring Continuous Capital Accumulation: "Climate Refugees" and "Climate Migrants"', *New Political Economy*, vol. 18, no. 3, 2013, pp. 337–63.

Bank–commissioned report on this subject emphasises, 'Migration is a proven development strategy pursued by agents to maximize their needs and values'.[60]

The neoliberal confluence of an apologia for migrant entrepreneurialism and a naïve humanism is perfectly illustrated by these two World Bank experts, whom it is worth citing at some length:

> Migrants of all kinds consistently display initiative in helping themselves. They are rarely hapless victims of circumstance; indeed, the very fact of their movement suggests that they take action to resolve problems. Labor migrants, for example, are hard working, seeking to maximize incomes to finance a better life for themselves and their children (irrespective of whether they are temporary or permanent migrants), to send money to families at home, and to help new migrants to overcome barriers to movement and settlement . . . The entrepreneurial endeavor of refugees and IDPs makes them a potentially important resource that can enhance the capacity of the hosting community to adapt to climate change.[61]

This new way of seeing things has the merit of casting off the more miserabilist aspects of neo-Malthusianism and securitarian doctrine. Clearly taking up a position within a neoliberal logic, it instead valorises individuals' capacity for action. These latter are no longer 'passive' victims but entrepreneurs of the self, able to spontaneously resolve whatever challenges present themselves and – in so doing – to increase the general well-being. The entrepreneurial qualities that these migrants have are exactly the ones that one can expect of the 'climatic' subject – capable of being flexible, of rebounding, of adapting not in one definitive way but rather to constantly changing conditions. In other words, here we again find the will to produce a resilient, flexible subject capable of profiting from the climate crisis.

---

60   Jon Barnett and M. Webber, 'Accommodating Migration to Promote Adaptation to Climate Change', World Bank Policy Research Working Paper, no. 5270, 2010, p. 26.
61   Ibid., p. 24.

'God helps those who help themselves' also seemed to be the conclusion of the 'Foresight' report commissioned by the British government, which helped redefine the terms of debate on climatic migration in the early 2010s.[62] Migrants were supposed to be capable of 'protecting themselves' from environmental risks, of 'extracting themselves' from difficult situations and of 'building a new life for themselves'.

Far from representing a tragic result of environmental crisis, 'it must be recognised that for many people [migration] is an important way of bringing themselves out of poverty and out of vulnerability to global environmental change.'[63]

In this logic, what actually presents a problem are the obstacles to this spontaneous adaptation – and thus these obstacles are what governments need to remove. In the first place, this means the barriers set up by the insufficient ability of the most vulnerable people to get moving. 'Caught in a trap', non-migrant populations are cornered into a space that is gradually becoming unlivable, and they are incapable of injecting the dose of mobility that would allow them to create a virtuous circle of migration, remittances, education, etc. On the contrary, they are building up deficits. In this view, the attentions of those intervening from the outside should concentrate on the 'immobile' rather than on migrants. Here we again come across the neo-Malthusian viewpoint which sees populations' immovability as a factor getting in the way of adaptation to climate change.

Nonetheless, it would be quite mistaken to think that this paradigm shift with regard to migrants implied any generalised opening of the borders! The likes of the IOM and the British government do not promote the free movement of people around the planet. This turns out to be a crucial point – and it is important to understand it properly. The encouragement of migration as an adaptation strategy takes its place within the context of an international-level 'migration

---

62 *Foresight, Migration and Global Environmental Change: Final Project Report*, London: Government Office for Science, 2011. For an analysis of this report, see Felli and Castree, 'Neoliberalising Adaptation to Environmental Change'.

63 Ibid., p. 182.

management' system which the IOM is trying to put in place.[64] Nothing would be worse, from this point of view, than migration being overly spontaneous, using illegal channels outside states' control:

> If left unmanaged, environmentally induced migration can have disastrous consequences, primarily for individuals and their communities. When a certain critical mass is reached, unmanaged migration can also have security implications for concerned countries with the potential to spill over across borders to neighbouring territories.[65]

This is necessary in order to defend migrants' 'human rights', of course, but also to ensure that migration will indeed be productive in character and contribute to overall economic accumulation. Even if the IOM now puts the emphasis on promoting mobility, as a means of adapting to global warming, migration management is still caught between the neoliberal and neoconservative paradigms just outlined.

The neoliberal approach to managing migration, as a strategy for adapting to climate change, also makes up part of the offensive against the sovereignty of Global South countries. Presented, as in the writings of Kaplan or Homer-Dixon, as failed states or 'ungoverned territories' incapable of managing their own populations and thus of managing the risks that arise within their failing structures (corruption, trafficking, uncontrolled emigration, terrorism, etc.), these states must be aided and assisted. This will be possible thanks to the outside

---

64   Ishan Ashutosh and Alison Mountz, 'Migration Management for the Benefit of Whom? Interrogating the Work of the International Organization for Migration', *Citizenship Studies*, vol. 15, no. 1, 2011, pp. 21–38 ; Martin Geiger and Antoine Pécoud (eds), *The Politics of International Migration Management*, Houndmills, UK: Palgrave Macmillan, 2010; Younès Ahouga, 'Le Tournant local du migration management: l'OIM et l'implication des autorités locales dans la gestion des migrations', paper presented at 'Les migrations à l'épreuve de la nouvelle gouvernance locale', 10–11 December 2015 at the Université de Poitiers.

65   *Compendium of IOM's Activities in Migration, Climate Change and the Environment*, Geneva: IOM, 2009, p. 39.

expertise supplied by international organisations, public development aid agencies, NGOs and private-sector multinationals, who together can provide structures of civil society that respect the human rights which these 'failed' sovereign states so lack.[66]

CLIMATE WARMING? HOW CONVENIENT

But if it is important to encourage the mobility of people vulnerable to global warming, then how, practically, should this be organised? The Zambezi Valley in Mozambique, one of the world's poorest countries, is populated by many hundreds of thousands of small peasants.[67] It is also subject to severe floods. The river level can vary rapidly, and when it does rise the water frequently invades fields and homes, as in 2007, when close to 100,000 people were temporarily displaced. After that, the Mozambican government threw itself into a vast relocation programme to move these populations out of this area, which experts had identified as an at-risk zone. Yet farmers have occupied this space and exploited its particular characteristics for many hundreds of years. They have developed cultivation techniques which allow them not only to prepare for floods, but also to take advantage of them – for they fertilise the soil. Floods appear to them less as disasters than as recurrent events inherent to this location.

The government nonetheless justified this displacement policy on the grounds that climate warming is increasing climate variability and would thus produce more frequent surges and floods. This viewpoint is shared by several international organisations, like the World Bank, which finance development projects in the region. In the name of adaptation to global warming, the government undertook the

---

66 David Chandler, *International Statebuilding*; see Hartmann, 'Rethinking Climate Refugees and Climate Conflict', pp. 239–42; and Felli, 'Managing Climate Insecurity'.

67 This presentation of the Mozambican case is drawn from Alex Arnall, 'A Climate of Control: Flooding, Displacement and Planned Resettlement in the Lower Zambezi River Valley, Mozambique', *Geographical Journal*, vol. 180, no. 2, 2014, pp. 141–50.

displacement of almost 140,000 peasants towards 'safer' habitations. And they were certainly safer – but also more distant from fertile land and water abundant in fish. This adaptation thus weakened or even destroyed the means of subsistence for part of the population – one that was already among the most vulnerable – at the same time creating fresh social problems. As for the peasant families who continued to inhabit the flood plain, the government considers them irresponsible and as deliberately choosing to expose themselves to climate risk. It has thus disassociated itself from any responsibility to help with rebuilding efforts should a disaster strike. Parallel to this decision, it has sworn off any further maintenance or development of the little public service infrastructure (hospitals, schools, roads) that exists in this area.

Yet the geographer Alex Arnall shows that while global warming does indeed have effects on climate variability, one cannot ascribe to it the main responsibility for the rising flood risks looming over this region. This instead owes rather more to the consequences of the building of hydroelectric dams upriver, notably the Cahora Bassa dam completed in 1976, which have added greater unpredictability to the fluctuations in the river level. Sporadically releasing large quantities of water in response to the demands of electricity production, these dams have greatly reduced peasants' previous capacity to anticipate the river's variability.

The real problem lies in the fact that the many uses of the river and its floodplain in this part of the lower Zambezi are in competition: there is hydroelectric production, but also commercial navigation on the river and, increasingly, an industrial agriculture (the result of the liberalisation policies carried out in the late 1980s). The sudden releases of water from the dams are thus also useful for the movement of boats and the irrigation of the great mechanised agricultural properties. But while they favour the interests of these commercial and export activities, they increase the vulnerability of the poor peasants and their production of food for consumption. As Arnall notes, 'Many of the investments ... responsible for displacement of small-scale farmers from the Lower Zambezi River valley – dams, industrial agriculture and navigation – represent important sources of revenue for the Mozambican government'.[68]

The systematic foregrounding of global warming and the need to adapt to it is thus heavily instrumentalised by the region's political-economic powers. Added to that, international funding – especially that from the World Bank – seeking to spur adaptation and resilience faced with climate change, seems to be allocated as a function of local political loyalties rather than real needs. In any case, this funding process disregards the participation of the people actually concerned in favour of 'the overriding goal of rapid disbursement'.[69] The flooding problems in the Zambezi Valley were without doubt serious, but according to a study by two development aid specialists, they did not figure among the priorities of the local authorities. As they note, 'It is hard to see any evidence of "country ownership" in this selection.'[70]

At the same time as the central political authorities are bypassed by the choices the World Bank makes, 'stakeholder participation' is subjected to a minimalist interpretation. In the name of urgent action, consultation with local populations – not to mention their active participation – is reduced to almost nothing. Through a raft of small and decentralised interventions (like programmes to restore the mangroves) the dispensers of international aid can thus rapidly disburse the allocated sums while allowing local barons to control access to financial resources. The problem Shankland and Chambote underline is that the World Bank programme 'allows decisions on major climate resilience investments (including tens of millions of dollars in loans that the people of Mozambique will be expected to repay) to be taken without broad civil society engagement or even public awareness'.[71]

The socio-economic conditions that produce poverty and increase insecurity in the means of subsistence are partly rendered

68   Ibid., p. 146.

69   Alex Shankland and Raul Chambote, 'Prioritising PPCR Investments in Mozambique: The Politics of "Country Ownership" and "Stakeholder Participation"', IDS Bulletin, vol. 42, no. 3, 2011, pp. 62–9.

70   Ibid., p. 65.

71   Ibid., p. 68.

invisible when we instead point to the responsibility of a distant process of global warming. The humanitarian concern to provide rapid protection to vulnerable populations conceals strategies to enhance economic accumulation and to valorise commercial activities whose use of space enters into competition with its use by more traditional forms of production. Submissive to the World Bank's desires, states use their power to create the conditions for neoliberal economic accumulation, aggravating the social inequalities and inequalities of power that prevail in the region. The population displacements carried out in the name of adaptation to climate change, as shown in this case, reinforce rather than challenge rising inequalities.

ADAPTING AGRICULTURAL WORKERS TO CLIMATE CHANGE – OR TO CAPITAL?

The fate of seasonal agricultural labourers in Turkey also illustrates how the imperative of adapting to climate change is instrumentalised on behalf of economic accumulation.[72] These workers are among the populations most vulnerable to the effects of global warming, precisely because of their working conditions: the lack of social protection, informal employment, direct exposure to bad weather, dangerous work, handling pesticides without protection, etc. Worse, the already arduous toil in the fields can quickly become intolerable as temperatures rise.

A slight fluctuation in agricultural production (exacerbated by climate variability) will drive landowners and farmers to offload their now-excessive workforce, without any compensation. In this context, seasonal labour migration represents a safety valve for both proprietors and the authorities: excess migrant workers can be rapidly cleared out without further costs to the capital invested in

---

72   The case study that follows is drawn from Ethemcan Turhan, Christos Zografos and Giorgos Kallis, 'Adaptation as Biopolitics: Why State Policies in Turkey Do Not Reduce the Vulnerability of Seasonal Agricultural Workers to Climate Change', *Global Environmental Change*, vol. 31, 2015, pp. 296–306.

agricultural production. The costs of 'social reproduction' are thus outsourced to other territories and other populations. The counterpart to this is, of course, the immense precarity – and thus the immense vulnerability – of these workers themselves, a genuine 'reserve army' who can be mobilised on demand, and who generally lack any alternative.

The district of Karatas, in Adana province, on the eastern Mediterranean coast of Turkey, receives close to 100,000 seasonal workers between February and October each year. There they cultivate cotton, tomatoes and different types of melons. The great bulk of these migrant workers are Kurds or Arabs. After the farming season, they head back southeast again to their homes. In the district studied by Ethemcan Turhan and his colleagues, the predicted impacts of global warming are particularly severe. Studies indicate a temperature rise of up to 6–7°C, inducing droughts and a decline in rainfall of nearly 25 per cent.[73]

But to understand the effects that global warming will have on these migrant workers, we have to understand why they are vulnerable to it. And this vulnerability is not climatic in origin, but rather has to do with social processes and institutions.

The existence of such major seasonal labour migration in the agricultural sector is to be explained in terms of the history of agrarian policies in Turkey in the latter half of the twentieth century. Applying the Marshall Plan, these policies favoured the 'modernisation' of agriculture – meaning, its capitalist concentration, to the detriment of small and middling farms. Unable to survive in an environment dominated by huge farms, a large part of the rural population was thus dispossessed of its land and its means of subsistence and had to migrate to the towns. The measures taken in the 1980s to liberalise the rural economy and promote free trade swept away the last remnants of the developmental state in Turkey. Confronted with the consequences of this rural exodus, the big landowners brought in a seasonal agricultural workforce from the east of the country (itself

73   Ibid., pp. 298–9.

consisting of peasants uprooted by the concentration of agricultural lands in a few hands). Part of the Kurdish population, forcibly displaced after confrontations with the Turkish army in the mid 1980s, was also forced to sell its labour-power to the landowners, under deplorable conditions, just to survive.

The fact that a large part of the seasonal workforce is of Kurdish origin further increases its vulnerability, putting it at the mercy of frequent police checks and the nationalist resentment of local landowners.

The vulnerability of the seasonal farmworkers subjects them to the decisions of landowners who work hand in hand with the Turkish state. Any fluctuation on the food market, or any fluctuation in the climate (less rainfall, rising temperatures) is immediately displaced onto the workers – the most flexible and vulnerable part of the production system – by cutting wages, making the work more intense and strenuous, or through layoffs. We see this in the case of watermelon growing, which represents an economic risk to producers that has grown with global warming. This is a lucrative crop, but one highly sensitive to the weather conditions in which it is produced. To reduce these economic risks, some producers turn towards cereals production, which is less sensitive to fluctuations of rainfall and temperature, but also far more mechanised. As a result, they lay off their seasonal workers.

The impacts of global warming thus affect social individuals and groups in a very different way, depending on their class. The positions of power which the landowners occupy allow them to offload the bulk of the damage onto more vulnerable populations. A policy of adaptation to climate change should, on the contrary, reduce social vulnerabilities in order to mitigate the effects of this calamity. But nothing of the sort is happening.

The adaptation plans prepared by the Turkish state (and implemented by the United Nations Development Programme) seek to maintain the circulation of a cheap and flexible workforce. This is necessary for maintaining profitability in the country's agricultural sector. So, while some projects for adaptation to global warming seek

to improve seasonal workers' living conditions during their employment, they do so while also making sure that these workers do, indeed,
leave at the end of their stay. That is, migration has to be continuous
– and any permanent settlement avoided – in order to reduce the
costs of social reproduction. Similarly, the very small measures taken
to improve housing and set up healthcare services seek above all to
monitor and control this moving population as best as possible, while
also making sure that it does not attempt to organise and claim rights.

When the Turkish state develops its adaptation policies for the
agricultural sector, it does not do this with a view to reducing the
vulnerability of seasonal workers. If that were the goal, setting up
social insurance and social protection would be an infinitely more
effective way of going about it. After all, this would make it possible to
'smooth out' climatic fluctuations from one year to the next; it would
also partly release workers from the grip of the intermediaries (who
negotiate their contracts and siphon off part of their wages) and landowners on the labour market. In this regard, Turhan and his colleagues
note that the state has 'the intention of maintaining labour costs low
under increasing climatic uncertainty, sustaining only the minimum
conditions necessary for labour reproduction.'[74]

But we know that the union struggles by agriculture workers are
setting the foundations for a transformation of the agribusiness sector
in the direction of low-carbon production, much more respectful of
the environment and of workers themselves. This is hugely important:
almost half the world's active workers are in agriculture, with more
than a third of them working for a wage (most of them women). Out
of all workers worldwide, these are the ones most at risk of an accident on the job, especially accidents connected to handling pesticides.
As Peter Rossman notes, the basic demands of workers in the agricultural sector – a job and a decent wage, as well as a safe and healthy
working environment – imply a profound transformation of the
global food system, towards less monoculture, less industrialised
production and a reduced use of chemical fertilisers. In short, the

74   Ibid., p. 304.

success of trade-union struggles could drive a transition toward an agro-ecology compatible with climatic limits.[75]

The Turkish state's interventions seek to do the opposite: to mould individuals as workers even more flexible than they are already. They do so by encouraging the temporary circulation of labour-power – to the benefit of landowners and to the detriment of the workers themselves. This reinforces the agro-industrial and capitalist orientation of production. The promotion of circular labour migration as an adaptive solution to climate change increases the reach of the market by flexibilising farmworkers' working conditions.

THE POLITICAL ECONOMY OF DISASTER

If it is undeniable that global warming alters the 'nature' in which human beings live, these transformations do not *automatically* lead to determinate social consequences. The long-posed equation between global warming and population displacement ought to be wielded with some caution – for overall transformations in the climate do not translate into the same outcomes for everyone. Wealth, property ownership, access to networks of solidarity or public aid, for instance, can either mitigate or amplify climatic transformations.

Ignorance of history – meaning, of the crystallisation over time of social, economic, gender and racial relations of forces – too often leads to a simplistic climate determinism being taken at face value. Even if it is endowed with the best of intentions, the acceptance of this determinism is a problem. When global warming is said to have 'caused' wars or migration – in Darfur or elsewhere – the real blame for these atrocities disappears from the analysis, as do the concrete and local means of fighting against the effects of global warming. Above all, the powers-that-be can thus use the shock of global warming to entrench their domination. Far from reducing populations'

75   Peter Rossman, 'Food Workers' Rights as a Path to a Low Carbon Agriculture', in Nora Räthzel and David Uzzell (eds), *Trade Unions in the Green Economy: Working for the Environment*, London: Routledge, 2013, pp. 58–64.

vulnerability, part of the adaptive solutions privileged in recent years have extended the market's sphere of influence, especially with regard to the workforce. Thus, when the likes of the IOM and the World Bank promote migration under the pretext that it represents an entrepreneurial strategy of adaptation to global warming, they push this mercantile reasoning to extremes. Since they do not advocate for a human right to freedom of movement, the promoters of this vision risk making migrants even more vulnerable, since they make the mobility of workers conditional on productive capital accumulation. As we have seen in the Mozambican and Turkish cases, the shock of global warming is used by the powerful forces locally to bolster big farmowners' domination over dispossessed workers. The chosen modalities of adaptation are not neutral: rather, they are the translation of a particular political economy – the political economy of capital. To ignore this political economy leads to the reduplication of the climate catastrophe with a political and social catastrophe.

Conclusion
Adapting Forever

As we come to the end of this study, only one thing is certain: if global warming is not rapidly brought under control, any talk about adapting to it is pointless. Even a few degrees of warming will have dramatic effects on ecosystems and on the possibility of inhabiting the planet over the coming centuries. The priority has to be the massive, rapid, socially just reduction of fossil fuel use. But this is an impossible task so long as the capitalist organisation of nature, necessarily oriented towards the exploitation of labour, energy and resources, continues to dominate.

The social, indigenous, peasant, feminist, and trade-union struggles today converging around the slogan of climate justice target the problem at its source. When they demand 'leave the oil in the ground', they point to the beginning of the only immediate possible solution.[1] Indeed, I do not see how it will be possible to avoid introducing measures to plan resource use in the effort to limit dangerous global warming.

Such measures could take two radically different forms. The first – the path chosen by governments since the 1997 Kyoto Protocol – consists of defining a maximum ceiling on greenhouse gas emissions and then allocating atmosphere-usage rights (pollution permits) as a function of the economic capacity to pay (the polluter-pays principle) and the extent of past pollution (the paid-polluter principle). As

---

1 Maxime Combes, *Sortons de l'âge des fossiles!: Manifeste pour la transition*, Paris: Seuil, 2015.

Polanyi would have seen, this solution contributes to the extension of the market. Most important, it follows the principle that 'everything must change so that everything can remain the same' – the famous line in Giuseppe Tomasi di Lampedusa's *Il Gattopardo* (The Leopard).

But there is an alternative. It would consist of subjecting the distribution of atmosphere-usage rights to the widest possible political debate – at the international level, certainly, but also at national, regional and local levels, and across economic sectors. The rights to use resources would, then, be shared out, by way of a democratic planning of the economy, in accordance with economic and social needs. They would be the result of deliberation – and not of the ability to pay. Such a solution implies a major transformation of social, economic and political relations of force. Only an intensification of the class struggle – and the conquest of state power by movements seeking to change how these states function and reduce the power of business – can force such far-reaching transformations.[2]

We have said that the climate question is, first and foremost, a political one – and it implies the proposal and then the enactment of a different way of producing the world. Calls to 'politicise' or 'repoliticise' the climate must, logically, result in an alternative political economy, based on solidarity, democracy and ecology. And this means a transformation of private property. Only this will make it possible to go beyond capitalism as the decisive organising principle of the fabric of our lives.

The implication of our perspective is that, from the outset, the climate question has been posed as an existential question for capitalism. Emerging in the context of the Cold War, from the 1970s on this was, inextricably, both a scientific and a political problem, especially in the United States. Climatologists and scientific institutions – combining with private foundations, sectors of the US state apparatus and 'enlightened reactionary' elites – sought to formulate responses

---

2 I will not elaborate further here (it is not the subject of this book), but Naomi Klein's book *This Changes Everything* describes the resistance being organised by climate justice movements and their perspectives.

that would allow for the continued pursuit of capitalist accumulation, even if at the price of great upheavals.[3] Up till the mid 1980s, these responses marginalised any possibility of reducing warming by cutting fossil fuel use. On the contrary, these various actors considered global warming (and especially climate variations) as a fait accompli to which society could respond by adapting.

Infused as they were by neo-Malthusianism, one first set of responses consisted of demanding an extension of (especially US) state power in order to promote the reduction of birth rates (in the Global South), a more assertive management of food reserves and some type of global planning of resource usage. Especially in the context of the 'green revolution', these first responses promised enhanced opportunities for the US agrifood industry and US agrifood researchers to ramp up their penetration into Third World countries, on the pretext of the challenges posed by climate change.

The neoliberal assault of the late 1970s and especially after Reagan's arrival in power in 1981 partly put the brakes on this visible extension of state action (albeit a state action that had acted in favour of US multinationals). This assault unleashed a Polanyian 'counter-counter-movement' of re-extending the market into spheres from which it had previously been excluded. Along with the attack on trade unions, environmental management was one of the first domains to suffer this re-extension of the market. In the eyes of the new elite fraction in power, the state's excessive involvement in developing alternative sources of energy, or in the direct management of food reserves, rapidly became suspect. US climate research programmes were dismantled or marginalised. But the climate question continued to be posed within the scientific-administrative elite.

In this new context, economists' voice would become decisive in reframing the climate problem: the climate was now portrayed as just

---

3   Thus I think that Bruno Latour portrays only part of the reality when he describes (in *Facing Gaia: Eight Lectures on the New Climatic Regime*, Cambridge: Polity Press, 2017, first lecture) a clash between climatologists who kept purely to the science and who were thus overwhelmed by climate sceptics who understood that they had to politicise the facts.

one of the many changes affecting society. Climate uncertainty echoed a number of other uncertainties, which all seemed to indicate the failure of public policies which sought to mould the future, and argued as well for the impossibility of planning. The neoliberal economists thus responded to this politics of uncertainty with the gospel of flexibility. The effects of climate change – as, in the last analysis, of all other changes – could be absorbed, surmounted, even put to good use, so long as individuals and institutions became more reactive, flexible, adaptable . . . And according to these economists, at least, the institutional mechanism that concentrated and enhanced all of these qualities was the market.

Despite the hopes raised for a decade or so (1997 to 2009), the international climate regime's incapacity to begin any effective reduction in greenhouse gas emissions would give fresh impulse to what we might call 'adaptive reason'. Yet this latter seeks, at best, to bring things back to normal (never mind the fact that the 'normal' is itself catastrophic for billions of people) and at worst to mobilise the climatic shock to extend the shock of the market into a multitude of domains (water management, insurance, international migration, etc.) and spaces – even if the Global South remains the favoured terrain for experimenting with these neoliberal counter-reforms. The theorists of resilience are right to say that social and ecological systems are now co-evolving on a planetary scale. But they fail to name the reason which animates this co-evolution: capital. For if the question today is how to adapt societies to the impacts of global warming, the reason this question is posed is that societies under capitalism's hegemony have sought to constantly adapt the biophysical world to their own needs – and through this process they have transformed themselves. The biophysical and human worlds' adaptation to capitalism's development into a global system signals our entrance into the times of what we can call the Great Adaptation.

How can we resist all this? As a minimum, by taking part in a new counter-movement in protection of society: against the extension of the market, against the consequences of global warming – and against the extension of the market which takes place in the name of protecting against the consequences of global warming! This counter-movement could begin by reducing the risks of disaster.[4] It would support policies which could reduce the vulnerability of the near-totality of the population. These would include universal social security (for old age, illness, accident, unemployment, etc.), more extensive infrastructure (transport, communications, applied research, etc.), subsidies (for energy, water, agriculture, etc.) and public services (education, healthcare, civic protection and emergency services, etc.) financed through solidarity taxes (and charges on the extraction of natural resources) or provided by public-owned companies and closely controlled by representative-democratic institutions.[5]

In short, this counter-movement implies a rupture, breaking from the neoliberal moment. It calls for the democratic extension of social protection, and a different way of organising the fabric of our lives. I believe that this environmentalist, democratic socialism constitutes the best hope for reducing climate catastrophe and maintaining freedom within a nature – irreducibly both biophysical and social – which is so complex and divided. To those who feared that this was the road to serfdom, I think it is worth remembering the words of Karl Polanyi, writing in 1944, in a time far more apocalyptic than our own:

> As long as [man] is true to his task of creating more abundant freedom for all, he need not fear that either power or planning will turn against him and destroy the freedom he is building by their instrumentality. This is the meaning of freedom in a complex society; it gives us all the certainty that we need.[6]

4 Ilan Kelman, Jean-Christophe Gaillard and Jessica Mercer, 'Climate Change's Role in Disaster Risk Reduction's Future: Beyond Vulnerability and Resilience', *International Journal of Disaster Risk Science*, vol. 6, no. 1, 2015, pp. 21–7.

5 This is the message of an important article by Jesse Ribot ('Cause and Response: Vulnerability and Climate in the Anthropocene', *Journal of Peasant Studies*, vol. 41, no. 5, 2014, p. 667–705), which proposes a dialectical approach to reducing vulnerability and the struggles for social emancipation – especially drawing on the works of the socialist-feminist philosopher Nancy Fraser as well as Karl Polanyi.

6 *The Great Transformation*, p. 268.

Index